Co

Economic Collapse

of

2006

by

Michael Wells Mandeville

Trends, Predictions, & Prognostications for 2004-2006 and beyond

Some say it has already begun... are you ready?

Are you ready to ride out the massive economic, political, and media power grabs by small elite empire builders who are destroying the incomes of labor and the middle class? Can you handle the rolling economic failures which are already drowning nations around the globe in deeper and deeper waves? Are you ready for the collapse of the value of the dollar and a huge increase in the price of all imports, including oil? Do you know what to do about the massive corruption and destruction of the U.S. government by greedy corporations? Are you prepared for the the gathering political storms which will wrack political parties, legislatures, and elections in the U.S. and around the world with polarization, fractionalization, controversies, and even violence over the Big Lies told by George Bush & his crooked cronies?

If you are ready, give this book to your neighbor!

Publisher's Signature

The Coming Economic Collapse of 2006

First Electronic/Paperback POD Edition: ISBN 09721051-9-0
A MetaSyn Media "Cornplanter Institute" Imprint
HC01, Box 890A
Black Canyon City, AZ 85324

Web: www.michaelmandeville.com/
Email: mwm@michaelmandeville.com

Retail Prices: US$18.00; BP 11.00; EURO 15.90
To order additional copies, order online www.michaelmandeville.com
Press and Trade inquiries: www.michaelmandeville.com

file begun 5:15 2 22 2003 as Transmutations
file transfer seven chapters on 9:24 5 8 2003 as ab_collapse2003_v1
file in final form on 8:25 49 pm 7 26 2003 as ab_collapse2003_v3
file edit - pricing and credits info added on 5:24 pm 8 1 2003
file edit – last typo debugging 10:08 pm 8 13 2003

Table Of Contents

Dedication

This book is dedicated to the perennial populist and progressive critics who have followed the same byways of consciousness made famous by Thomas Paine. Great and glorious acts of divine patriotism and sacrifice to higher cause must be accomplished all over again during the coming few decades. Greater legends than have ever been told are yet to be born. May God bless each and every one of the coming revolutionary apostles of justice for the family of humankind.

Credits

Thanks to the support of the Phoenix Quest Fellowship and subscribers of the Earth Changes Bulletin, this material was journalized week by week and then was transmuted into this book.

Thanks to all those who fed back ideas and crtiticism, especially David VanAmburg, most of whose thoughts and tips made it into this book.

Thanks as well to Kimberbeth Simmons who gave me the idea for the cover and encouraging me to continue breathing fire into the story.

And thanks indeed to Pam Wiseman for a last minute edit to clean out all the typos and weird glitches.

Cover: The graphs shown on the front and back cover are authentic charts of the daily closing average of the Dow Jones Industrials. The 1920's data was quoted from the book: "The Dow Jones Averages, 1885-1995", edited by Phyllis S. Pierce, Irwin Professional Publications, 1996. The 1990's and 2000+ data was supplied by the data engine at Yahoo Finance.

Section One

What This Book Is About

TOP SECRET - PROGRESSIVE EYES ONLY

Falling Through The Rabbit Hole

A funny thing happened to me and a whole lot of other people as I was on my way to drafting up a 40 page report on the "*25 Year Economic Depression Cycle*" which is due to hit in 2006 or 2007. I wanted to carefully outline the time frame of this cycle for my friends and my subscribers to the Earth Changes Bulletin. I also wanted to alert them to the last running of the Bulls in this cycle and how to gracefully liquidate all entanglements, careers, and paper investments which might not survive a real economic depression.

It seemed like an easy task. The future looked pretty straightforward during the Spring of 2002. After all, the conservatives were now in charge of the U.S.A. Everything was supposed to become simpler and easier. Beyond getting tough on terrorist organizations and throwing some more money to their pals, the Republicans were not likely to make any moves which would change much. And Bush Junior was sure to stimulate the economy more aggressively than Bush Senior, he knew his re-election depended upon it. Give a little more time to smash a few more Al Qaeda Cells and for trickle down money to flow through the economy, surely everything would start to roll again.

The only thing people needed to know is that the stock markets would bubble up again for about three years and then the economy would collapse to complete another down wave in the same clock-like cycle the economy had been on since the early 19th century.

Easy, I could finish the task during a few weekends by drawing together some of the material I had previously written for a prior book and add a few contemporary numbers. So I went to bed in early Spring, glad that things seemed to be moving along rather well.

When I woke up in the middle of Summer, I realized my worst nightmare had been realized. There was no limit to Ashcroft's increasingly ghoulish demands for unhindered police power, Bush Junior really was trying to instigate an invasion of Iraq, and most of his team of administrators were members of a cabal of "neo-cons" who really did advocate going after all of the "Axis of Evil" countries with military force, and a number of other Muslim states as well, if they did not kowtow to U.S. "dicTAT" given with the same sort of tact used on Afghanistan and Saddam.

Already by mid-Summer, reaction patterns and international political polarization were spiraling out of control. The economy was headed South once more and the stock market had just begun blinking its "Tilt" sign. Before I could even grab a cup of coffee, it was clear that the next few years were not going to be your average type years. Forget about the 40 page report. There was going to be nothing simple about summarizing the coming collapse, if "summarizing' it was even possible.

Indeed, it was impossible to get a handle on how events might now unfold until after the invasion of Iraq. The domestic and international political scene kept getting more and more surreal, a Dali-esque tapestry of weird over-reactions, snafus, lies, a U.S. Secretary of State and a British Prime Minister who attempted to prove their case with a term paper plagiarized from a student, more lies, and bizarre insults resounding through an echo-chamber of hyperventilating right wing kooks who suddenly could be found toning their sorcery buzzwords under every rock and in every TV and radio broadcast.

Seemed like Central Casting was trolling the very bottom of the barrel for this "new normal", as CNN news mavens loved to call it. How could things get more strange, I wondered, and THAT was a mistake because then I noticed that CNN's motto had become "The Most Trusted Name In News", displayed as a title in mega-sized type which grandiosely filled the screen to the accompaniment of triumphal music flourish, the sort of music for rapt attention and stiff upper lip which fascists love.

That's when I knew that it was all over. Anybody who has to sell that hard, has a problem. We had all fallen down the rabbit hole and we were on the way to the Mad Hatter's Tea Party. We were late, late, very late, the White Rabbit kept exclaiming nearly everyday from a variety of podiums, for a most important date in Baghdad.

Let's see, where am I? I had gone to bed in the late Spring as war was winding down in Afghanistan to write a simple report over the weekend, I had awakened in the middle of Summer in the middle of a nightmare which had just begun and no matter what I did I could make no sense out of the Mad Hatter's party. All I could do is wait for a while until a new "new normal" stabilized out while the dust settled down in liberated Iraq.

As I waited and watched, it came to me that both myself and large numbers of other people were truly beginning to awaken to a profoundly new understanding of realities about the U.S., its history, its politics, its economics, and the new "Imperial Faction" which had emerged in control of its government. We were awakening to a profoundly troubled path on which America's new leadership circles and Mass Media were gleefully leading us to a "future" where "everything was now different", not to mention revoltingly fascistic. The pundits of the "new normal" had hardly a passing notice to the 5 billion people they had profoundly alienated for another generation.

Finally, this May 2003, nearly a year later, things appeared finally stable enough and certain trajectories seemed well enough "set" to risk defining the most likely probabilities and major contingency factors for the next several years. Actions and reactions seemed locked in enough so that commentary would not be totally obsolete the next day. The next Six Sections thus finally emerged to spell out what is underway and where it is going. Maybe.

Definitely this is a maybe. We have such power-crazed, neurotic stuffed egos in charge of the U.S., it is not easy trying to figure out the weird ways in which the U.S. is going to insult the planet next and evoke the Devil only knows what kind of hostile reactions.

How lucky we are! We get to wake up every morning now to see Lord Acton's famous dictum proven anew in ever more undeniably obvious ways: power corrupts, absolute power corrupts absolutely. America 2003 has a real bad case, an epidemic running now throughout vast economic and political realms. And everyone, EVERYONE just wants to pretend that it isn't quite real.

Are you ready to deal with the massive economic, political, and media power grabs by small elites, the rolling economic failures which are already sweeping the globe in deeper and deeper waves, the collapse of the international value of the dollar, the massive corruption of the U.S. government by greedy corporations, and the gathering political storms which will wrack political parties, legislatures, and elections in the U.S. and around the world with polarization, fractionalization, controversies, and even violence?

That's where we are going. Ready? Ah well, I didn't think so. After all, we are a little bit ahead of the story. To bring you up to speed, the remainder of this section summarizes this report and the basic story line for the next several years.

How & Why This Book Was Written

During the late Spring of 2002, I was upbeat about how many events were unfolding for the World and for the U.S. Events appeared to be turning progressively positive while New York healed from the September 11 attacks on the Twin Towers. Most nations, even those with frosty relationships with the U.S. government, had successfully united behind an unprecedented worldwide campaign to destroy Al Qaeda and other terrorist networks.

The Bush Administration seemed to be "in synch" with the world. Bad historical situations, some of which had been festering for a generation, were beginning to turn around. Afghanistan was setting up its first enlightened government, many terrorist attack plots had been interrupted, Al Qaeda was in flight everywhere, scores of Homeland Defense programs were being implemented to prevent future terrorist attacks, and signs of revival in the economy seemed evident. Even a solution to 50 years of controversy and violence over Palestine seemed on the verge of peaceful settlement. George Bush and Colin Powell had recently committed the U.S. to host an international peace conference to solve the Palestinian conflict once and for all in the Fall of 2002 with the full weight of the international community.

For a brief while that late Spring, one could even suppose that the U.S. economy might begin to recover its growth trajectory, making possible new initiatives and opportunities to propel the economy past the dot-com bust. By historical standards a recovery was already due and the Federal Reserve had progressively reduced interest rates nearly to all time lows. Republicans in Congress were promising extremely liberal "write-off" allowances for small business capital purchases, which everyone expected to widely increase demand for automobiles, computers, and communication technology. Surely, it seemed, the breeze must now pick up and sail the economy onward while people picked up their lives which had been so rudely interrupted by Osama Bin Laden's followers.

I personally began to prepare to publish information about a *"25 Year Economic Depression Cycle"* which I had found while reading some of Edgar Cayce's prognostications about history. (See "*Return of the Phoenix Book Two: The Great Breakup*" by MW Mandeville; MetaSyn Media 2000; p. 181.)

According to the "cycle", another depression or severe recession was due in 2006 or 2007. Shouldn't the existence of such a cycle be published and made available to people everywhere, I asked myself? The crash of the dot-com bubble in the midst of this relatively mild but lengthy recession was bad enough. Shouldn't everybody have some advance notice about the next crash, which, looking at the historical charts, was virtually inevitable?

After considerable reading of the previous two hundred years of history, I had been able to confirm the long term cycle, which Cayce had described as 24 or 25 years in length. I proved it up as part of my work during the late 1990's in writing the story of Edgar Cayce's life and his predictions in a trilogy of books titled "*Return of the Phoenix*". Without fail, all but one of the worst depressions and recessions in American history fell right on the 25 year interval which Cayce had described. The one which didn't during the 1830's was merely off by three years.

I tucked the definition of the cycle into the second book of the "*Return of the Phoenix*". In that context, the cycle attracted scant notice from people who deal with such things, much as I thought would be the case. For the most part, academicians and experts in today's world are so lost in secular materialism and narrow topical horizons, they can't bear the thought of dealing with people who can accurately and consistently make "unguessable" predictions about the future while being wrong less than 1 time out of 10. Many or most academics, having examined very little of the universe, are nevertheless quite certain that no one has or can do such a thing.

Without much audience interest in the prediction, I decided to bide my time until the time was closer. I figured that people would become more interested in the high probability of a major crash in many aspects of the economy if the event were looming less than four years in front of them. Thusly, in the late Spring of 2002, my thoughts began to return to thinking about that cycle and how the economic events might unfold.

I began thinking about the cycle with the thought that another running of the bulls would begin by the end of the year (2002). The Bulls would lift up the stock markets while growing more jobs, making 2003 an up year on all levels. I envisioned the coming bull run as possibly similar to the Bull Run of 1926-1929. Surely once more all "ships" would rise through nearly four years of steady growth, very much like during the famous "Roaring 20's" rise just before the gruesome crash which heralded the Great Depression. Then, beginning perhaps in 2006, the crash would bottom out the economy over a two or three year period, reducing many stocks to $.10 on the dollar, similar to the magnitude of the collapse of Enron. Enron, I reasoned, was probably an example of what would happen eventually in 2007 to perhaps dozens of large companies which had vastly over-extended themselves.

It was an interesting challenge to present the case for such hard times. But I felt I had all the experience I needed to sift through the data and the strange, obscure fantasy worlds of the economics profession. I had spent some nine years in college studying complex systems, philosophy, history, economics, especially economic history, and politics. Such background made me comfortable separating fact from academic jargon fantasies while cruising through both the internet and many varied news beats.

During the 1980's I had loafed with my academic knowledge for several years while reviewing two to three dozen business plans a week which had been submitted by businesses and corporations who were requesting funding from a venture capitalist. This taught me how to do feasibility assessments on the fly in a variety of industries for both technology applications and "business expansion projections". In reading these piles of "business plans", I encountered every exaggeration, deceit, slight of hand, and scam ever thought of and then some.

Even before then, I had became thoroughly grounded in the politics of international energy trade and some of the many efforts to create alternative energy technology since the inventions of Tesla. One year the State of California even sponsored me to show over one million Californians how to produce alcohol fuel in several county fairs and the state fair.

Later, I began to really work some of my systems knowledge while writing some 25 business plans over a period of many years for various clients in a variety of industries. This gave me an even deeper appreciation of cashflow, industries, markets, economic fluctuations, and banking and finance institutions.

All this business and technology background taught me how to see the obvious and how to see the differences between facts, fantasies, fads, and fallacies. Reducing provable claims out of projections and speculations was a never ending quest to determine the feasibility of the technology application and the likelihood of a positive economic outcome.

During the 1990's I applied this same hard-edged analytic skepticism to produce an elaborate hard-scientific validation of Edgar Cayce's predictions. Fair and square, I determined that he was 92%

accurate on general economic matters and I published the results in the "*Return of the Phoenix*".

This was an extremely interesting finding which I frankly did not expect. I thought originally he might be a little over chance, maybe 60% at best. Finding his statements so provably "probable" changed some of my own paradigms about the nature of the world. I had to take his tips seriously and they pointed to realities which were often intellectually uncomfortable for the most commonly accepted "conventions" of "acceptable" viewpoints, both secular and religious.

The more I followed up his tips related to the geology of the Earth, the more I found unsuspected scientific connections between many unexplained geological and astronomical phenomena. Based on the various tips which panned out, I could put 2 and 2 together about how the cosmic mechanics of the Earth, Moon, Sun work to produce tectonic plate movements, earthquakes, volcanoes, El Nino and La Nina syndromes, even Global Warming!

These connections took a lot of mystique out of the environmental changes which are underway in the Earth. The connections work through the Earth's wobble in space (known as Chandler's Wobble) which are caused by minute variations in the orbits of the Moon and the Earth and by the shifting distribution of mass on the surface of the Earth, primarily the water. The motions caused by the wobble produce most of the geophysical and geological trends, changes, and events in the Earth, including the latest trends of "Global Warming".

Without the slightest intention of doing so, purely as a hobby, I ended up defining the discovery and proof of a new scientific paradigm for geology and geophysics which I have named Vortex Tectonics. It explains the basics of how "the whole earth" works as a system through long geological ages.

So, using the powerful tools and links on the internet, I began to monitor "the whole earth" to look for correlations with the theories and principles of vortex tectonics. I began to communicate a raw feed of what I was discovering on the internet through an email newsletter which I call the Earth Changes Bulletin. The most exciting aspect of this new view of the world is that it actually works to predict and explain what is happening. Bad weather periods, for instance, can be plotted out years in advance!

Based on many tips sent in by subscribers, I began to expand my monitoring to include sunspots and planetary alignments. By connecting planetary alignments with solar activity and Earth's weather in various locations during the past two years, we (my subscribers and I) mutually discovered and verified that certain planetary alignments produce sunspots which create the major storms and extremes in Earth's weather patterns.

This was so interesting that we began to look at how the high sunspot counts affect human reactions, war, the economy, and the course of history. I discovered that this has been well studied, though not widely known, since at least the beginning of the 20th century. Research by Professor A.L. Tchijevsky, a Russian scientist, was presented in a paper to the American Meteorological Society at Philadelphia in the late 19th century. He showed beyond question that wars were closely correlated to the rise and fall of the numbers of sunspots. If one diligently plots out the 11.5 year sunspot cycle for the past few hundred years, one can easily see that nearly all major wars or major political changes occur during or very very close to the high peaks in the numbers. (See the Sunspot Cycle Charts in the Appendix).

Exactly why this is so cannot be explained very well but the correlation is very good. The simple explanation is that the ionic and magnetic influx affect physical, emotional, and mental functioning, but the real explanations get quite fuzzy beyond that.

Every 11.5 years sunspot numbers range from a monthly low daily average of about 10 sunspots to peaks of monthly average daily counts as high as 150 or so. Daily highs can reach nearly 500, but usually only during the Peak periods. The lows persist for about four years, the peaks above 100 persist for about four years as well, leaving about two to four years of counts in between the two extremes.

This is a HUGE dynamic shift for a cycle and it includes a wide variety of sometimes awesome activity, such as sudden explosions, which can easily exceed the collective size of the four inner planets. Since the advent of SOHO satellite technology and other scientific investments in solar monitoring which have been made during the past 20 years, scientists have learned that the peak sunspot periods are accompanied with huge solar flares, storms, vast magnetic storms, and "mass ejections" of the Sun's atmosphere which often bombard the Earth with tremendous amounts of electrons and protons moving at speeds up to about one tenth the speed of light in vast streams of "solar wind".

Occasionally, protons even arrive from a solar mass ejection or explosion in as little as 30 minutes. That's matter and that's fast, carrying prodigious unit energy!

Regardless of the speed, the ionic influx of the solar wind makes quite an impact on the Earth's atmosphere and this is what causes the extreme weather patterns. The ionic influx during the sunspot cycle peak also causes intense emotional reactions, hardened attitudes, and quite effectively propels humans to increased violence, war, and enough paranoia about society to draw back investing and spending to cause recessions and depressions.

Many economic downturns are at least partially connected with the sunspot peaks. Many are found to begin just after the middle of the peak and they often endure until the sunspots return to very low levels, much as appears will be the case with the current recession (of 2000-2003).

Considering time as structured by sunspot peaks, the Palestinian Infiltada was right on time. The new wave of suicide bombings began just as Solar Cycle 23 rose to a peak above 100 in 1999. Osama Bin Laden was right on time as well. The Twin Towers were hit virtually in the middle of the peak for Solar Cycle 23.

Americans appear to really like the backside of the sunspot cycle peak, at the point where the cycle is still high but is beginning very slowly to decline. That is the point when the U.S. entered World War II and when Lyndon Johnson sharply escalated the Vietnam War. The Bush Administration was right on "solar" time with the invasion of Iraq. It commenced just as Solar Cycle 23 Peak was beginning to slide down, just at about the same part of the curve when the U.S. invaded Iraq during the 1991 gulf war. Charts which display these relationships with the sunspot peaks, as well as with the economy may be found just after Section 7.

Doubtless the economy is also influenced by the solar cycles but exactly how is much less clear. Looking at the Great Depression, it is clear that a huge spike in sunspot numbers gave the 1929 DJI "bubble" the enthusiasm to reach 389 before beginning it's long slide to about 40. And of course, the high peak of the current sunspot cycle was the moment of truth for the dot-com bubble. In a sense, the bubble kept inflating until the ionic influx burst it past the ability of any logic to hold it together. One can make other similar connections, but they are not nearly as consistent as sticking with "*Cayce's 25 Year Economic Depression Cycle*".

All these vortexian and solar connections bring us to a wonderful place of unity and linkage between vast phenomena and tons of data about them. Doubtless, even as the mystics have always exclaimed, there is a pervasive real unity of planets with sunspots with cycles with energy with seasons with weather with oceans with people with wars with economy with magnetism with quakes with the wobbles in the spin with El Nino with Global Warming, and does it end anywhere?

Using these connections, I now regularly predict global weather periods on the internet through the "*Earth Changes Bulletin*" and the "*Earth Changes 2003 Almanac*". During the past two years we had one "false alarm" for bad weather but we have not failed so far to anticipate the major drought patterns, the flood periods, the extreme storm fronts, and the most likely windows for both relatively calm weather and really calamitous weather on most continents simultaneously. I forecasted, for instance, this most recent May 2003 window which brought the sudden outbreak of hundreds of very destructive tornadoes. I described the two weeks as full of "extreme chaos and freak extremes in the weather". This I knew would result from a huge surge in sunspots which the alignment between Mercury and the Earth (Mercury between the Sun and the Earth) would create at the beginning of May.

By focusing at solar and global levels, which weather people do not do, I would see the interlinking of many things which produce the main "big patterns". Necessarily, the timing and the location of our forecasts in the "*Earth Changes Bulletin*" are somewhat sloppy because nobody has made supercomputer modeling available yet for this global perspective. But even so, our subscribers on the internet know where to get fairly decent "big picture" forecasts which are as reliable as the local and continental weather forecasts. The "*Earth Changes Bulletin*" puts out forecasts of extreme weather periods with at least 75% accuracy.

These ideas will take some time to become accepted by experts. Only a few scientists feel comfortable tackling such a range of subjects as part of one interacting, unified system. They prefer to take one small slice at a time. That is convenient, and it is sometimes the only way a topic can be tackled, but that never provides a grasp of the "living universe", only a piece off a cold corpse.

The Bulletins have achieved at least as good a rate of accuracy for political and economic affairs as in the weather/flood/drought predictions. I ignored these fields until the Spring of 2002, even though I was deluged by subscriber requests for more info related to these areas. I had long ago concluded that secular materialism and the entire political spectrum of right vs. left was a doomed paradigm, a field mainly suitable for interminable arguments which never get resolved.

So I personally preferred to look in the spiritual realm for answers. The old idea that it all starts with you, that you have to be the change you want to see in the world, seemed to be as good a place to start as any, perhaps it was indeed the only place you can start. So that puts the focus on your own spiritual beingness, awareness, and values.

But clearly, in a world of 6 billion people, many of whom are threatening extreme forms of violence on others, one's own sphere is not enough. Eventually one cannot avoid the issue of how to join one's sphere of activity with others to accomplish higher ends than burying the corpses of countless victims of violence. This issue can be ignored for a time, but it cannot ultimately be denied. Those who do ignore it, become eventually the victims, not the winners.

The Mad Hatter's Tea Party

Willful denial of the political and economic spectrum began to end for myself and large numbers of other people in early summer 2002. It became perfectly and embarrassingly obvious that, even at the height of paramount success in uniting the world to fight the War On Terrorism and create peace in Palestine, the Bush Administration took a wrong exit off the "Peace Freeway". Instead of choreographing an international peace conference over Palestine, which Bush and Company had solemnly promised to do, various politicians, administrators, and Mass Media personalities began an intense propaganda campaign to rationalize invading Iraq.

Within a few weeks, the intense beating of Mass Broadcast Media war drums was creating a cacophony of discord around the world. As well, everything economic turned sour. The recovery stalled, if there ever was one, and the depth of corporate corruption was revealed to be wide, deep, and very stinky. The Dow Jones Industrial Averages began to sink even more, the dollar lost value, bankruptcies continued to mount, and the aviation and tourism industry went from the basement to the sub-basement.

At the time, I was intensely watching how the sunspots were driving the droughts and were causing worldwide damage in the agricultural industries on all continents. I was concerned that the depression waves from all of this rural contraction would drive the recession far longer and cause more disruption and bankruptcy.

With the advent of the talk about invading Iraq, I concluded that the sunspot activity was defocusing government leaders. They seemed to be making the same grandiose expansions in their ambitions which had brought down many nations in the past. During a Sunspot Cycle Peak, at a point near the end of the peak, very close to the favorite spot where Americans take up wars, Hitler grandiosely began his invasion agenda in Poland and ignited World War II. Perhaps, fed by too much ionic influence from the Sun, Hitler's arrogance waxed larger than normal to ultimately destroy both himself and several tens of millions of people. I wondered if the Bush Administration, influenced by Sunspot Cycle 23, could be making a similar type of historical error.

So it was that I began to monitor the primary variables in the economic landscape, including the value of the dollar, the DJI, movement of the economy, political reactions to both the economic indicators and the moves of the Bush Administration related to Iraq and the War on Terror. With the help of some of the Earth Changes subscribers, I began to hone in on making both short and long term predictions about the course of international events.

I left sunspots behind on this quest, and relied most heavily on systems science and objective analysis of the actual interactions between nations and factions. I employed the same sort of classical analytic arts about politics which Aristotle taught and which the CIA once used during the 1960's and 1970's (perhaps they still do). Without doubt our calls were generally way ahead of the curve. We articulated a few speculative scenarios, one or more of which usually occurred. And we made many specific calls which came to fruition. Our calls were very close, most of the time.

And that is why this Special Report has come about. Rather than just write weekly projections as we come closer and closer to the next downturn of the "*25 Year Economic Depression Cycle*", I have put together projections which attempt to encompass our movement through the entire next six years or so.

Putting it together is quite a bit like solving a who-dunnit-murder-mystery. Based on the time frame defined by the "*25 Year Cycle*", we can compare with previous cycles and crashes. From this we can learn to "see" what the real dynamics are, what the typical political reactions are likely to be, and thus what the nearly inevitable outcomes must be.

We begin point by point with what has been happening and what the current conditions are. We take seriously the arguments of French President Chirac, Russian President Putin, and other foreign leaders. We look carefully at where current events are likely to lead us in many areas, both economically and politically. From this we compare with the previous cycles to find out how things are likely to end up in the coming crash. We look for the new beginnings, as taught by history, which most likely will be part of the recovery process.

Hopefully, the comparisons provide some serious perspective on how we can personally enter and exit the major economic and political shifts which must flow between now and, say, 2008. Ultimately, the comparisons, especially those as found in Section 7, provide a "crib sheet" for solving problems, making personal decisions, and for buying and selling during these shifts. In a sense, this is a personal business "plan" which outlines some key strategies for sailing though some very tough, ugly, stormy times...times which may be more like 1929-1936 than anything since.

How The Logic Unfolds In This Report

A strong image of our most likely future begins with the "Time Frame" given to us by Cayce's highly accurate "*25 Year Economic Depression Cycle*". With greater than 92% odds, this cycle will give us a depression or a very serious recession in 2006 or 2007. Everything else falls into place before and after and the dynamic outcomes of some of the current trends suddenly become obvious.

Because of the absolute importance of this prediction, this cycle is rigorously laid out in **Section Two**. How it was found is described, why it is credible is discussed in great detail, and the historical proof for it is outlined. This provides us with a basic guideline, a criterion for sorting through all the other possibilities and potentials about our future.

From this Time Frame we know with 92% certainty that the economy will not grow in 2006, nor will it limp in 2006. We know that it will flatline or contract in 2006 and it will bottom out no later than 2007. From this, we know an immense number of other things about the inevitable dynamics of the economy (and the political arena) during 2006 and 2007, as well as for prior years and succeeding years.

Doubtless then, the "Time Frame" composes our destiny "funnel". It is an inevitable period of exceptional stress which we will all experience together. It is a time of extreme compression into which everything even now is moving. I say "funnel" because of the narrowing of all things, all horizons, all aspirations, all plans, all wealth, all production. Human affairs will contract and many things of the past will disappear for good, as they usually do during such periods, before new horizons can be glimpsed at the other end of the funnel.

The most important question about this prediction is how severe will this depression or recession be? Will it be like Reagan's rather severe recession? Or like Eisenhower's rather average recession? Or like Hoover's Great Depression.

Section Six *(The Dollar Panics)* will provide some ammunition for answering this question. As more and more analysts are beginning to argue, all may hinge on the value of the "petrodollar" relative to other currencies, especially the euro. Given the severe structural problems in the U.S. economy, which run a gamut of issues, most notably the Balance of Payments and the massive "jobs flight" into China, India, and other countries, there is a lot of reason to suppose that the value of the petrodollar will plummet and that the coming economic downturn will then become more like the Great (Worldwide) Depression of the early 1930's.

Some of the basic issues related to these structural problems are dealt with in **Section Six**. The petrodollar's background is painted for the past 30 to 40 years to provide wide perspective. Against this background, one can see that there are many negative trends creating a bigger and bigger "float" of bills which are coming due. These have and will continue to generate a much bigger bubble than the dot-com bubble.

The potential depth and breadth of this shift should not be underestimated. The Collapse of 2006 or 2007 is, more probable than not, more than just another recession or depression. The collapse is most likely synchronous with the "moment of truth" for making major structural changes in the economies of the world, especially for the dollar and the U.S. economy. The trends of the past 25 years are progressively forcing big changes on everyone in every nation and most international analysts readily concur that a lot of "re-adjustments" are due.

In particular, the huge float of dollars seems to spell big trouble. Looking at the larger historical scene, one could argue that the dot-com bubble was just a minor bit of froth on an even bigger bubble which is yet to burst. One could continue to argue that the dollar is the huge bubble which must burst and that if it does, the whole stock market, especially the high tech sector, must crash to maybe $.30, $.20, even $.10 on the 2003 value of the dollar.

One could look at the Enron crash as an example of what will ripple through a large number of corporations, the "Fortune 500". One can even see Microsoft as vulnerable to a major crash in value, down towards the level of its actual tangible cash assets. One can argue this, not on the supposition that its books are cooked, but simply on the basis that its time is over, that its products are widely distributed and during a major depression, people will change dramatically how they spend money. Who needs more Microsoft software? Who really needs the latest versions? A lot of high technology companies may face this same issue. Huge numbers of people may suddenly lose jobs in the computer related industries.

A major part of the dynamic which we must confront during the crash of 2006 is "globalism". The huge historical shifts which are underway are changing national economies into parts of a "global" economy. This shift has progressively created massive career dislocations, trade imbalances, islands of stagnation, untenable debt and credit relationships, and severe environmental damage. It has also brought unparalleled consolidations of capital and wealth by power seekers and brokers who manipulate the changes to serve the agendas of small ultra-rich elites which are often times adverse to the best interest of large numbers of people. These manipulations are raising the levels of anger in many if not all nations throughout the world.

As during the 1930's, the systemic adjustments to resolve many of these issues may be socially and politically cataclysmic. Extreme political crisis may result from any of many sources of hysteria, especially over the issue of jobs. Economic hardship from sudden economic "adjustments" which is shared by large numbers of people never fails to generate widespread mass reactions. These are easily fanned into violence, progressive social anarchy, extreme radicalism and coups, even violent revolutions among societies with too many Marie Antoinette's in charge. And these reactions may generate, as they have throughout history, even more wars and revolutions.

The sort of dismal events which occurred in the 1930's could unfold but they are not necessary, they may not even be probable in most of the world. People with decent historical perspectives may arise to hold in check the impulses and reactivity of the worst morons among us. The operative words are "may arise". America has far too many Marie Antoinette's in charge of far too much for anyone's good.

For the most part, specifically how these adjustments will take place is not foreseeable, probably by the grace of God. It is enough to be aware that there are a large range of possibilities. Accordingly, the most important thing to do is to watch developments carefully to see what is blowing in the wind. The second most important thing to do is to learn how to lean with other people in the right direction to effect change in the direction of governments and economic policies.

Given this unknowability, perhaps the best strategy is to be prepared for anything, from the most typical of recessions to the most dire of meltdowns. **Section Seven** provides a core strategy for surviving the downturn, no matter what comes, and a lot of tips for positioning oneself to profit (constructively and fairly) during the next growth phase. A major part of this strategy ought to be creating maximum liquidity beforehand along with a personal survival base which insures food and shelter in a rural area. Another part of this core strategy is knowing how to back the right political party with the right industrial policy which can deliver a new era of economic expansion with the most appropriate "recovery" industries.

Before taking up the structural issues, one should pour a very strong cup of coffee, perhaps make a new pot, and reflect on the major political changes which have been accelerating since the 911 attack. For this, we mainly need to tune into the present and the past year or so to tabulate what we need to watch. This reflection on the present begins in

Section 3 (The Global Political Shift) and continues on through **Section 4** (The Global Media Shift) and **Section 5** (The Global Economic Shift).

The most important dynamic shift to catch is the Spring 2002 Surprise, which is when we fell through the Rabbit Hole and Bush's cabal launched the U.S. on its invasion drive. The policies, strategies, and tactics of Bush and Company have induced many major changes among the nations of the world. Some were good but since Spring 2002 a great many of these are not good changes. Some of the changes are profoundly accelerating the cumulative negative effects of the structural problems related to trade and the value of the dollar, all of which are moving South. This is now happening in so many ways at such a rapid clip, events could easily begin to spiral out of anyone's control. In fact, that is likely what will happen to create the Collapse of 2006.

But, adverse trends could go the other way, depending...on elections, policies, personalities, foreign invasions, and so on. What are the main trends and reaction patterns to watch? Do the existing dynamics tend to an amelioration of the conflicts and problems or do they tend to a worsening of conditions, polarization, deadlock, and violence? How are the mass reactions likely to play out? Can we see peace, or more invasions and wars? Terrorism and more fascism? International deadlock in the United Nations? Or genuine alliance building? Is economic expansion and growth in the offing, or continued confusion? There are many other questions equally as important.

The simplest truth is that some trends and conditions in some respects are improving while others in many ways are deteriorating rapidly. How the balance will play out is the crux of the matter. Will the positive reaction patterns and uplifting trends overcome the damage to our economies and societies, or will the negatives swamp us into a deep cesspool of paralyzing conflict? It is a serious but open question. The answers are blowing in the wind and we will attempt to capture them as best we can for the moment in this Report and through time as it moves along in the "*Earth Changes Bulletin*".

Sections 3 and 4 pull no punches about where things currently stand. The negative political and economic trends induced by Bush and Company are the worst the world has seen since the nuclear arms race during the Cold War. So long as Bush and his Faction (they call themselves neo-conservatives) are able to manipulate Republicans and maintain control of the U.S. government, the negative trends are likely to deepen and worsen and produce greater and greater polarization.

Some of the major negative trends are:

(1) An extremely strong domestic polarization which has divided the U.S. right down the middle;

(2) Global polarization which has unified about 90% of non-American public opinion against many American policies;

(3) Devastating loss of the credibility of the statements of the White House and the U.S. government throughout the world; the U.S. President, unlike any President in living memory, is widely perceived around the world to be a lying buffoon whose incompetence will eventually catch up with him;

(4) Widespread loss of credibility of the American Mass Broadcast Media (TV and cable network news systems); much of the electronic media appear to have incessantly and manipulatively beat the war drums to push Americans to support the invasion of Iraq; this is extolled as a great American triumph of virtue and heroism but most of the civilized world sees the entire affair as just another sorry episode of American jingoism;

(5) A Nazi-like concentration of unparalleled police powers in the U.S. which are far beyond constitutional bounds and any rational means of over-sight, the virtual destruction of traditional American values of checks and balances for equal justice;

(6) The astonishing consolidation of wealth and power accompanied by unabashed, blatant corruption of political and legal functions (such as the give-away of billion dollar contracts to a handful of oil industry insiders who collectively are very close campaign contributors to Bush and Company);

(7) The splintering of both major U.S. political parties – collapse of cohesion within both parties and the likelihood that traditional conservatives will bolt the Republican party while liberals bolt away from the centrists in the Democratic party; a probable radical re-alignment of political parties and the formation of agendas which lock the U.S. in stalemate;

(8) Increasing economic disintegration for the lives of greater numbers of people.

On the other side of the coin there are some good countervailing trends which may come to the fore in various ways and offset some of the negatives. Here are some of the positives:

(1) Eurasian cooperation is increasing: most profoundly for the future of the world, some of the major countries of Eurasia (Germany, France, Russia) have made the conscious decision to form a concerted political bloc in opposition to Bush to counter-balance the U.S. and outweigh its economic strength;

(2) European common defense strategy is being rethought by leaders who want a common defense force which is European, not connected with American forces;

(3) Internet-news media has bloomed worldwide in ten thousand flowers through providers which are determined to offer powerful means to antidote the inadequacies and political manipulations of the Mass Hysteria Broadcast Media; it is getting more and more difficult to get away with lies at any level of government;

(4) New global wave mass communication media, independently owned, is beginning to appear in video and audio formats through cable, satellite, fiber, and internet channels; these are "crystallizing" worldly perspectives which will progressively supercede parochial nationalistic attitudes;

(5) A rapidly rising populist movement in the U.S. is beginning to stir with considerable vigor; the movement is using the great power of the internet to transcend all previous limitations of time and space; doubtless much more will be heard from "the people" during the next few years than has been heard during the previous 30 years;

(6) Spiritual leaders of the major religious traditions are beginning, like the Ents in "The Lord of the Rings", to awaken to the realization that the world aches for them to transcend the limitations and divisions of the past to provide moral authority which heals and liberates and uplifts the human species; increasing activity and cooperation between the various religions shall surely change the political environments.

These are the main political changes for which we need to prepare and position ourselves, both personally, and collectively. Can we say how most of these trends will all play out through to, say, 2008? For most if not all of them, not really. The best we can do is watch them with an eye towards their ability to impact what we are trying to do to survive the economic changes.

Domestically, if the positive forces of progressive populism and reform prevail, the coming collapse will serve mainly as a powerfully bitter medicine which cures some nasty conditions in the body as a prelude to a rapid rebound after 2007 or 2008. In this circumstance, those who are poised for rebound in the basic household markets and the new industrial technologies will be greatly advantaged. They should grow rapidly in the period 2008-2012.

If the negative forces are prevailing, especially through bitter divisions and stalemate in the national government over economic industrial policy and issues related to "globalism", security, and foreign policy on war and peace issues, all outcomes will be highly unknowable. The chaotic dynamics of mob panics will become stronger and stronger.

We will recognize this situation if and when the times are full of demonstrations, marches, and rumors of war or invasions as a nearly constant, imminent threat. Everything will seem to come under a cloud of fear, insecurity, confusion. Strange panics and manias such as surrounded the Tarot Sniper, the Anthrax Letters, and SARS, will seem to be ever constant. We may even see repeat plays on these three themes, history rarely runs without copycats.

Under this situation, it will be both the worst and the best of times. One will be able to see the glass half full of opportunity for personal and collective transformation, or one will be able to see the glass half empty. Most outcomes will be highly uncertain but deep change will be a certainty. If so, creating liquidity and maintaining resources for the most basic modes of survival will be the most suitable strategy for large numbers of people. Those who are liquid when the crash finally bankrupts the corporations and many of the wealthy elites who control them (who to a great extent are debt ridden), will find they can pick up distressed properties for a fraction on the dollar, maybe for as little as $.10 on the dollar. Do you want a small farm? Save cash now, buy later.

Next on our list to dig into, as if we did not have enough already, is the Global Media Shift. Since the Mass Media, especially the Mass Broadcast Media, are profoundly manipulating every aspect of our politics, thoughts, awareness, reactions, and votes, we need with Eagle

Eyes to watch as these Sorcerer's cast their electronic spells over the consciousness of humanity.

This Sorcery is taken up in **Section 5**. The years 2001-2003 have astonishingly revealed very clear patterns of mass manipulation of American consciousness by the editorial class in the Mass Broadcast Media (collectively the network news output of TV networks and cable channels). They have maintained a systematic campaign through nearly two years of time to manipulate people into blindly supporting imperial policies, international aggression and invasions, and the imposition of extreme extra-constitutional domestic restrictions.

They have unabashedly supported the corrupt consolidation of wealth and power along with a ceaseless barrage of near-bigotry from vapidly stupid right-wing commentators whose "intellectual" market cannot by any stretch of the imagination be greater than 10 to 20 per cent.

Journalists in the broadcast industry used to worry about the ethics of TV news. They once seemed to care how far they could go in tele "casting" the news without outright being in the business of manipulation or as serving as merely a politician's lapdog. Now, a thousand examples could be cited for the past year to make the case that most of them no longer give a damn. It appears that lapdogs can have very cushy careers.

Accordingly, it is very important to understand that you cannot rely on a single word which is disseminated on TV, not even the video and pictures, which quite often are staged, phony presentations, NOT REALLY NEWS AT ALL. It is far more realistic for the average person to consider TV as merely a knee-jerk propaganda machine of, for, and by Wall Street, Big Money interests, and the right wing extremists who have taken over the Republican party. When a Mass Media channel brags they are the most trusted name in news, you know you should notch up your distrust to another level.

Against these power brokers, who have been far more supportive of extreme concentrations of wealth, fascism, and imperialism than anything else, it is time to unabashedly bash back. And indeed, a mighty reaction is forming, both domestically and internationally.

A New Media is arising on the internet and in many nations around the world. It is bashing back. It is highly supportive of an expanded consciousness of what is really happening in the world, in Iraq, in the United Nations, in history, even how the Bush Administration and the American Mass Broadcast Media are manipulating political opinion with continual broadcast of hyperbole, unproven claims, outright deceitful statements, juvenile chauvinism, occasionally even the expression of bigoted attitudes which technically constitute "hate" if not 'hate crimes".

Using the internet news media, we can tune into the main dynamics of all of the trends to carefully watch and assess them with far greater power and reach than can be imagined. And we can also tune into how the power elites through Broadcast Sorcery are trying to manipulate people to react to the trends. With a little effort, if we look carefully, with a little help from our friends, we can chase the Big Lie right on out of at least our personal lives.

The process begins in this report with the outlines in the following six sections. The process continues on the internet where we share, even as we have been doing for the past year through the Earth Changes Bulletin, important information unavailable in print which updates us about how the trends are likely to impinge on us.

Section Two

The Coming Economic Collapse Of 2006

According To Edgar Cayce

Can we predict the economic charts for the next few years and see the coming upturns and downturns? Can we use such knowledge to effectively position ourselves rationally to profit on the upturn and ride out the next coming downturn, which many are already expecting to be a severe depression? Absolutely YES!

Naturally, we have to leave academic sciences and professional economists in the dust. They have no way to predict the next six months let alone the next four years. Instead we must find a more ancient method, such as exemplified by Edgar Cayce, a world famous clairvoyant during the 1920's and 30's who was eagerly sought after by other famous people, including Thomas Edison, Woodrew Wilson, and Franklin Delano Roosevelt.

While reclined in a deep meditative trance, Edgar Cayce indirectly predicted in 1931 that the world would suffer another deep recession or even depression in 2006 or 2007. Cayce gave this "indirect prediction" as part of an highly interesting clairvoyant description of a long term "economic depression cycle" which is 24-25 years in length. By tracing this cycle out, Cayce told his listeners, you can find and anticipate the major depressions and deep recessions. From this you can knowingly predict that the next major economic crisis will come in 2006/2007.

Cayce gave many other predictions about the millennial period of 1998-2001 and the first decade of Century 21. These are mainly dramatic predictions of major religious, political, economic, social, and geological changes which Cayce typically referred to as sweeping ***"changes in the earth which must come again"***. By calling them ***"changes which must come again"***, Cayce gave a clue to his readers to compare this current era of time (at the beginning of Century 21) with catastrophic periods in earlier times which had caused major changes in history and the destiny of nations.

Cayce also made hundreds of predictions about his own times and the Cold War era which began very soon after his death. Nearly all of these predictions were objectively fulfilled, earning him the highest accuracy rate which has ever been scientifically assigned to a psychic, a phenomenal accuracy of 84% to 96%, depending upon the type of prediction he made. (See the *"Return of the Phoenix"* by MW Mandeville for a thorough scientific analysis of Edgar Cayce's predictions.)

What does this tell us about the coming economic collapse of 2006/2007? To begin with, it tells us that it is well worth while taking Cayce's predictions very seriously. If one had followed Cayce's economic predictions and invested in accordance with his guidelines during the 1930's, '40's, and ever since then, one could easily have amassed a multi-billion dollar fortune. That is compelling reason to sit up and take notice and plan accordingly.

Here is how Cayce achieved his phenomenal accuracy and provided his clients (and even his readers in this generation) the foreknowledge to create multi-billion dollar fortunes. He correctly described 5 to 30 years in advance all of the fundamental economic trends, the increases and decreases in interest rates, gold and silver prices, the gains and losses in the stock markets, the coming technologies, which industries would flourish, areas which would develop more than others, and many other variables, essentially all of the major developments of the 20th century.

It was as if Cayce tuned into the "book of the history of the 20th century" and read it all out days, weeks, months, years, and even several decades ahead of time for his clients. He made most of these predictions with precision far beyond the abilities of economists, futurists, and science fiction speculators.

He described to his investor clients the technologies and industries which would create the most economic growth both before and after World War II. He correctly forecasted, based on his 24-25 economic depression cycle, all of the major economic downturns between 1931 through to this time (2003). He made several other important predictions for this current era (1998-2012, including most importantly the nature of international and domestic conflict in the world. His predictions describe quite well the general "religious" nature of the current conflict which has caused the War On Terrorism.

Perhaps most spectacularly, Cayce predicted over a year in advance (in 1928) the Dow Jones Industrial Average Point (near 385) at which the stock market would begin to collapse in 1929. Not satisfied to just predict the point, Cayce also described the conditions under which the stock market and the economy would collapse and the duration of the Great Depression (no recovery in the stock market for five years after the collapse, he warned them more than a year in advance of the collapse). And, although his more conservative clients were often not happy with what he told them, Cayce also predicted the major steps which had to be taken to forestall revolution and lift the world from the Great Depression (he advocated dumping gold as the arbitrar of currency, eliminating high tariffs, bringing all labor unions into the councils of government, creating social security, and many other measures, all of which Franklin Delano Roosevelt eventually set in motion).

No other clairvoyant in recorded history has such an astounding track record which spans such a wide range of topics and so long an era of years. So when Cayce tells us that the "*Economic Depression Cycle*" at work in our society will create another one of the worst of economic downturns in 2006/2007, one should listen intently. It can be taken as certain that those who take the information seriously have 93% odds (Cayce's rate of accuracy for long term economic trends) of being well prepared for the worst. Obviously, those who do not heed the information may be positioning themselves to take major losses. Compared to the coming contraction in 2006-2007, the recent stock bubble collapse in 2001 and the long winded recession of 2001-2003 are a relatively mild downturn (just a resetting of the economic sails).

With highly accurate foreknowledge of this coming depression, people can prepare and shape their destiny to meet this severe shock. Investors can preposition themselves to avoid losses and those who like credit card spending will know they need to reduce their debt balances.

This foreknowledge will help mightily in properly raising one's sails during the next expansionary period of the economy. There may very well be another "bull run" of the stock markets during 2004 and 2005 before the next "break" and slump into the next depression, which will bottom out in 2006 or 2007. By using good prognostication of the likely trends and growth areas, knowledgeable investors may position themselves to profit quite well as the DJI bounces back up above 10,000, perhaps even up above 11,000 before the sudden collapse.

Or, if there is not another Bull Run, you will know how to compare the signals of this current era with the signals which preceded the past depressions. This will help you liquidate what you need to let go so that you can reposition yourself to take up better opportunities after the coming crash.

Like Cayce's clients of 70-80 years ago, individuals at all levels of income and wealth can devise appropriate strategies for maximizing their capital gains during the period 2003-2006. Most importantly, investors can use the same TYPE of "positioning strategy" which Cayce outlined for his 1928 clients. They will know how to watch the signs for when they have to liquidate real estate and stock portfolios to ride out the depression with the least damage.

About two years prior to the Great Crash of 1929, Cayce began to teach his clients how to liquidate their investments in the old declining industries and position their assets into the new leading growth sectors of the economy. As in most major downturns, some industries will never recover as much as others. New ones will suddenly become the leading growth sectors, such as personal computers did after the Reagan Recession of 1981/82. A few thousand dollars in the industries which Cayce specified for 1934 would have led to a very comfortable retirement during the 1950's. Similarly, ten thousand dollars invested into the PC industry in 1981 would probably have resulted in a multimillion dollar fortune.

To gain a secure foundation in using this foreknowledge, one must understand Cayce's prediction, its historical accuracy, and a sampling of some of the other amazing predictions Cayce made about the economy and the unfolding of the future. This information unfolds in the following pages of this chapter. For assessing these predictions with others made by Cayce, the exact scientific accuracy rates for Cayce's clairvoyance is also summarized below.

To fully use Cayce's foreknowledge to its best advantage, one must understand how the coming depression of 2006/2007 will result from existing economic circumstances and dynamics. What is now changing and will come to change even more during the next few years must be seen and understood in order to "make the right moves".

To help readers devise the right strategies for the next three years, the next few chapters provide a discussion of the current major geopolitical (Section Three) and geoeconomic (Section Five) dynamics which are determining or conditioning all other economic factors and are causing the ups and downs of our lives. By weaving your strategy with the "up" variables while avoiding the "down" variables, the maximum growth potential is most likely to be found.

To keep this information continually updated and constantly expanding for better understanding of the dynamics in play, this document is integrated with an online subscriber database at the *"Earth Changes Almanac & Calendar"* under the title of this report.

The Long Economic Depression Cycle

In 1931, Cayce described a 24/25 year cycle which accounts for all of the worst economic recessions and depressions in North America. Because of the large number of minor recessions which have appeared, this 24/25 year cycle had never been noticed prior to Mandeville's compilation of Cayce's objective historical and science predictions in the *"Return of the Phoenix"*. (This section is taken almost word for word from Book Two of Mandeville's *"Return of the Phoenix"* trilogy).

Cayce's description of the cycle may have lain without notice for so long because of the obtuse grammar with begins the description:

> *(Q) When will be the peak of the next period of general business prosperity, and the bottom of the next period of general business depression?* **(C) As the cycles come and go, these may be judged much for the future as they have been in the past. There is, as has been seen, (those general periods - going back, as we find) the greatest DEPRESSION as arose from ORDINARY causes (Those in the present are OUT of the ordinary, for they are the COMBINATIONS of both wrath, oppression and SIN), in the one nearer akin, as we find, in 1907. Hence a cycle seen - this, then, of the greater next depression, would be another cycle, or a seven (7), which comprises - as is seen - near to twenty-four to twenty-five (24 to 25) years. This, you see, makes the combination of peak made - oppression or depression, also of business.**
>
> 09/08/31 282-003 /10
> ECONOMIC PREDICTION FULFILLED

Cayce seems to be saying that there is a seven year cycle which, in combination with other factors, makes a longer term cycle of 24 to 25 years. What Cayce is saying about a seven year cycle is not clear. But those who have looked at Vortex Tectonics (see *"Vortex Tectonics"* by Michael Mandeville, 2001) may think that this seven year cycle is the 6.5 to 7 year cycle which is in Chandler's Wobble.

Despite this fuzzy 7 year item, Cayce's description of a 24-25 year cycle of economic depressions is pretty clear. Is it a good one? Will we find them in the historical records? The answer is an unequivocal yes. (For additional data on this subject see Mager's *"The Kondratieff Waves"* and the London Economist Newspaper's "*World Business Cycles*".)

The year Cayce references is 1907, which is not a bad pick. But 1908 is better. A recessionary period started in 1907 and the GDP actually contracted in 1908 by 8%, which made this period a true "depression". The next year it was up. Thus 1908 is the bottom year for this depression. For all of the other contractions, it is much easier to spot the exact "bottom year" of the depression or recession in the historical profiles than the exact start. One only needs to keep in mind that the economic downturn usually begins during the preceding year..

There were two other contractions in the GDP between 1909 and 1930, but they were each smaller, which seems to correspond to Cayce's commentary. If we add or subtract 25 years to 1908 we get the following table. Since 1784 the cycle only once has missed a beat. It was late by 3.5 years for the 1833 downturn. Since 1858, the cycle hasn't missed a beat.

During most of these downturns the economy actually contracted, i.e. had a bottom year in which the total value of production in the economy (GDP) was smaller than the preceding year. The main deviants for the U.S. economy were the 1958 and 1982 recessions. These two recessions, though worse than all of the other recessions during the cold war period, did not result in an actual contraction of the overall economy, just a slow down to almost no growth.

Table 204: Cayce 24/25 Year Economic Depression Cycle

Interval in years	Cayce Bottom	Recorded Bottom	Note
25	1783	1784	Recession
25	1808	1808	Depression
25	1833	1837	Bank Crisis
25	1858	1858	Depression
25	1883	1883	Depression
25	1908	1908	Depression
25	1933	1933	Great Depression
24	1958	1958	Eisenhower Recession
25	1982	1982	Reagan Recession
25	*2007*		*Next Recession or Depression*

This Cayce "Economic Depression Cycle" predicts a notably severe economic downturn every 24-25 years right on the money. This cycle does not predict all of the downturns, but it definitely predicts the bottoms of the worst ones quite accurately.

These economic bad years are not by any means all of the depressions or recessions which have been recorded in the U.S. There were several other recessionary periods. Economists, Prognosticators, and Market Tooters have defined many different cycles, and there are many schools of thought on the issues of economic cycles. A 24 year cycle has been identified, as well as 54, 50, 48 (Kondratieff Waves), 7, and 3 year cycles. None of them work perfectly in predicting economic events in any given year and all of the long range models by the economists and statisticians are quite "sloppy" indeed. They can vary by several years, which makes them worthless for tightly managed bull run investment strategies.

The DJI "385" Crash of 1929 Prediction

Cayce was not only adept in predicting the economic depression cycle, he was equally skilled in being able to foresee the exact "breaking" in economic trends and stock market values. Beginning in 1925, Cayce began to outline the shape of the bull run in the New York Stock Market and the exact dynamics of the October 1929 collapse. In 1928 he made what is perhaps the most spectacular prediction of all time, that the market would begin a vast downturn when the Dow Jones Industrial Average was "near 385". He advised his clients that this downturn would liquidate the world as it was and things would change drastically thereafter.

The "near 385" crash prediction is typical in the way in which some of Cayce's most profound predictions have lain so obscure. At first blush it reads as a prediction about Montgomery Ward stock going up to 385, along with a prediction about rail stocks going down. As such the prediction was wrong. Montgomery Ward stock was generally around 100 during the year before and not much better than 150 at the time of the prediction and thereafter.

Yet when one studies the prediction to analyze the portion which refers to the rails stocks, one can see that it was in essence an interpretation of dream images for Morton Blumenthal. Almost all of Cayce's stock predictions (thousands of them) for Blumenthal were "interpretations" which Cayce made of Blumenthal's dreams. Blumenthal dreamed Montgomery Ward, but Cayce interpreted that as a symbol for the entire stock market. It was not Montgomery Ward, it was the DJI which would go up to ***"near 385"*** and then begin a long decline.

When read properly, the prediction contained three specific criteria which would demonstrate when the Blumenthals should know that the end of the bull market game had come. The prophecy, besides being the most historically stunning, was the most specific and accurate prediction in the entire collection (*"Cayce/Davis Collection"*) of Cayce's 14,000 plus clairvoyant readings. Curiously no one else had ever remarked upon it.

The actual predictions related to the crash of 1929 were spelled out complexly in dozens of "readings" over a span of many years. Here is a summary of Cayce's bull run and crash prediction. In early 1925, Cayce predicted ***"adverse forces"*** would hit the financial world in 1929 after a long bull run. Through the next four years he advised his clients how to profit from the great bull run which led up to the collapse. He consistently described, a month to several months in advance, the shape of the curve of the bull run at several junctures, describing the breaks and the downs as well as the ups. These were genuine predictions based upon a "clairvoyant" ability, there was no way to consistently guess the up and down calls which Cayce made. Based on these calls, some of Cayce's clients made (in total) millions of dollars.

Cayce issued a second warning in May 1927 that the bull run would not last and that his clients had two years (to May of 1929) to fulfill their goals, else they would have to start all over again in 1934, which indirectly described the span of the great depression in North America. In late 1928 he correctly advised his clients that political controversies among the bankers and major economic interests would create increasing instabilities in the stock market over the issue of the easy "credit" which was fueling the speculative buying of stocks. He went on to correctly describe in advance the instabilities in the Dow Jones Industrials and the timing of the political controversies during 1929.

In probably the most chilling, spooky prediction in the entire Cayce/Davis Collection, Cayce described, one year in advance, the downturn of the stock market just prior to the crash. He described, within one point of the Dow Jones Industrials, when the market would turn, ***"near 385"***, which in fact occurred on September 1, 1929 at 386. He even described correctly other ***"signals and criterions"*** when this would occur and mysteriously named the number of the day in which the ***"indications"*** of the break would show, near a 26.

Two weeks later, in the middle of October, 1928, Cayce gave a definitive specification of exactly <u>what</u> the turning point in the stock market would be. It is so remarkably precise, it tends to blow one's paradigms. We have four fundamentally important statements in this complex prediction:

> (1) 358 is the first marker number, **THE SIGNAL**, when ***"changes in others may be expected"***, (2) the bull run will continue to near 385 then a major downtown will come, (3) the main break is when the downturn of the DJ Rails begins, and (4) that will be the time to refocus all activities to connect with new information through Cayce which will yield power and affluence.
>
> see(10/19/28 reading 900-412, paragraph 2)
> The Cayce/Davis Collection on CDROM

On August 2, 1929, the DJI went high at 358, **Cayce's signal number**, but closed at 353. The DJI got substantially choppier through August,, see-sawing between 337 and 380 (at which it ended the month), while the rails went from 173 on August 1 to 188 on August 30, only slightly choppy. Obviously the railroad interests were making a concerted last stand, but it ultimately proved no better than Custer's last.

On September 1, 1929, the DJI went high at 386 and the rails went high at 190, both closing slightly lower. That was the break in the market, **EXACTLY**. The Blumenthals and all of Cayce's other clients had ten days to liquidate everything at roughly that high plateau. Plenty of time to ride the next 10 years on the gravy train for all concerned.

They didn't pay attention!!! The industrials and the rails both slide slowly through September. By the end of September 1929 , the market was still extraordinarily high at 349, but DJ Rails were down from a high of 190 to 175 on a clear consistent downward trend. For the first time during the long bull run, the DJ Rails finished the month lower than the start of the month. Rails finally had broken the bull run and the ***"criterion"*** was whistling **down** the tracks.

In October the DJI slide to 300 on October 24 (so-called Black Thursday) and then crashed definitively on the 28th by falling to 260 in one day. The DJI finally fell all the way to 41 by July 1932, the DJ Rails to 13, marking the bottom of the depression when for all intents and purposes almost the entire North American continent was collectively bankrupt.

Cayce tried to warn them. But few of his clients acted appropriately. No one could believe that Titanic America was going to be broadsided and sunk by the icebergs of history. In the end the moneyed class had won, nothing. Only illusions, sleight of hand, political connections, mutual conspiracy, and licenses from the U.S. Treasury to print money permitted most of the moneyed class to survive.

After spelling out the 385 signals near the end of 1928, Cayce advised his clients in March 1929 to play ultra safe and get ready for a bear market with a very long gradual slide in values. He predicted that the market would continue to advance for ***"many moons"*** with increasing instability, then break for a long slide. In April 1929, Cayce added a reasonably good, simplified analysis of the political controversy which had deadlocked control of monetary policy, and warned that a great financial disturbance would result from the deadlock.

Just prior to the crash in October, Cayce defined a master strategy for a new era. He pointed to a fundamental change in the nature of the economy, away from the railroads into the new field of mass communication through radio, telephone, and telegraph. Cayce suggested that stocks could be bought in these fields but only on the basis of conservative price-earnings ratios (which would not have been possible until 1931). These, he told them, would be BOTH the safest, if bought correctly, AND the very best speculative bets because these new industries would grow the fastest. He finished the crash era predictions in October 1929 with an ominous statement about shadows appearing on the present horizon, heralding the dawn of a new social understanding.

The Great Depression

Second only to the stock market "break" of 1929, some of Cayce's most clairvoyant "five-star" predictions were on the span of the great depression. He accurately called all aspects of the Great Depression era, its beginning, its progress, the timing of changes, developments which would create the turn-around, its end, and the eventually return of a booming civilian economy after World War II (after his death). He made approximately 60 specific objective economic predictions which forecast all of the trends well in advance. His vision over the age is uncanny, it is as if he was able to read from a book which had already been written, though he maintained that the future was never pre-determined. Consistently, several years in advance he laid out the entire course of the American economy, even giving dates, years, comparative numbers, and specific industry forecasts. Only four of them failed, giving Cayce an economic score of 93%.

His readings, though often grim with moral reminders about the potential outcomes of selfish and parochial thinking, were consistently upbeat about the abilities of people to transcend the Great Depression and rebuild a better foundation for the world. On the one hand he exhorted people to apply the principles of the "God Game", which in Cayce's theology were very much about expressing love for God in the form of unselfishly creating justice, equality, and true respect for all differences between races and classes. He urged people repeatedly to use these principles to set in motion the new driving force for the world economy, that doing what was right for all people equally would solve all problems in the long run. Some of his commentary militantly preached accommodating the labor movement and the working man in politics and corporations for a great "***leveling***" of opportunity and income in the "***new order of conditions***".

Cayce warned over and over again that the only obstacles to peace and prosperity were imperialism, racism, religious zealotry, and the greed of self-serving elites. Has anything changed?

Yet despite his strong anti-war stance, and well before the demands of the global war (WW2) gave impetus to them, Cayce pinpointed the industries which would expand and eventually become dominant in the post-war world. He had a remarkable eye for the growth sectors of the age: aviation, telecommunication, radio, television, electronics, and plastics. Here are some of his calls.

The Great Depression

Second only to the stock market "break" of 1929, one of Cayce's most clairvoyant hits was on the great depression. He grimly predicted in December 1930 that it would get worse in 1931 and then get worse again, finally readjusting for recovery in 1934 and 1935. In 1930, nobody was capable of making such a guess. At the time, most people still thought of the downtown as merely a recession, which would be over sometime in 1931.

Recovery From The Great Depression

In August 1932 Cayce correctly predicted that the beginning of an economic turn around had started. He predicted the recovery of the economy ***by '35, '36, '38.*** It was a very good projection, far more upbeat and successful than all of the economic pundits of his day. By August 1932, Cayce was even predicting which areas of the country would rebound first, despite the fact that anarchy was growing all over the country and no one believed that there would be a change anytime soon. He forecasted that the commodities of gold, silver, steel, and construction materials would be the first to reflect a visible sign of improvement, which they did. He predicted that the southern and southwestern states would show the first signs of recovery, which in fact they did. Throughout the remainder of the '30's Cayce predicted several months to a year or more in advance how some of the various industries performed. In November 1933 he went much further and predicted successfully that the principles of Roosevelt's New Deal programs would become part of the permanent structure of the U.S. even though many of the specific acts and procedures would have to be changed, as indeed was later made necessary by reactionary Supreme Court decisions in 1935 and 1936. In response to dozens of questions by businessmen, Cayce made many successful predictions from 1935 through to 1943 about the implementation of various government programs and policies.

Recovery Reform Measures

In August 1932 Cayce predicted that the recovery would not begin another cycle of speculation and inflation and in fact it took nine years before inflation raised its head. He foresaw one to three years in advance the radical reforms under the New Deal which would change the nature of money, banking, and securities practices. Cayce was quite clear that going off the gold standard (partially) was a major factor in the turn-around. In May 1933 he affirmed that the revolutionary measures which the Roosevelt administration had imposed on money, gold, credits, and banks had laid the foundation for a sound recovery.

Gold, Silver, & Currency

Cayce was equally good on the trends of gold, inflation, interest, money and tariffs. He was never wrong. There isn't an econometric statistician on the planet who can match the record Cayce achieved psychically during the 1930's. He foresaw one to three years in advance the radical reforms under the New Deal which would change the nature of money, banking, and securities practices.

Throughout the 1930's and early 1940's he successfully predicted, several months to several years in advance, the ups and downs of interest rates, bond prices and yields, the periods of stability and flux in the inflation rate, and the behavior of gold and money in and out of the U.S. Less precisely, he also predicted the stabilization of exchange rates for foreign currencies. He also predicted that the use of silver and gold to back currency would eventually end.

Most spectacularly Cayce predicted that the price of gold would remain pegged at $35 per ounce of gold, which indeed it remained until 1971. Cayce told us in 1936 that the gold standard would eventually wane over a course of 50 to 60 years but he over-shot the final abandonment of the gold standard in 1971 by the U.S. (and thus the world) by 15 years. Even so, it was decently good prediction, equal to the one he made about international currency stabilization. He correctly could see that no international currency would emerge, for a long, long time, and gave no prediction about the successful creation of one.

In October 1936, Cayce foresaw that gold would not begin to flow out of the U.S. as a result of the improving economic condition in Europe, mainly because the capital coming into the U.S. was fleeing the coming of World War II, which Cayce named "***the world on fire***" in 1936.

Cayce claimed that changing the tariff law was essential for the recovery of international trade and he predicted that it would undergo change and become a tool of recovery. Indeed it did change radically and become one of the enduring cornerstones of U.S. foreign policy as the reciprocal trade act (most-favored-nation). With it, the Democrats destroyed the ability of special interests to dictate tariffs and employed it to negotiate mutual reductions with other nations to expand U.S. exports.

International Investments In The United States

He foresaw that greater and greater amounts of foreign capital would come into the U.S. for parking in a safe haven and as a growing demand for war materials. In April 1936 Cayce predicted that America would become what Roosevelt later termed "the arsenal of democracy": ***"America must remain - as it were - the balance of power in not only the money forces of the world but ... for the manning of ... destructive forces ... powering other nations with implements of destruction of every nature, these will show for a greater and greater volume being carried on in the United States."***

In April 1938 during a recessionary period in the midst of a lot of doom and gloom about international events, Cayce gave another one of his amazing long range five-star predictions: ***for some years yet ... more and more investments will be made of foreign capital in America ... gradually making for a period - which must come, of course, before '41 or 2 - when there will be the GREATER BOOM in the United States that it has known!*** The U.S. economy boomed an astounding

16.25% in 1941 and 15.43% in 1942 to deliver the greatest boom which has ever been known in the American economy, before and since. This boom, of course, resulted from the sudden huge demand for heavy war goods to fight world war II.

In June 1944 Cayce predicted, in his last economic prediction, that great economic changes would come after the war ended, but it would come after two or three years of economic disturbances. The war ended in 1945 and the civilian economy began to bloom in 1948.

Cayce's Clairvoyance In Science & Technology

Cayce provided many descriptions of the workings of the world and the cosmos. Many of them seemed implausible in his day but during the past 50 years none of Cayce's comments about science or the natural principles of how the cosmos works has been disproved. Many have been proven and more and more of his ideas have come to form the basis of modern science.

Cayce described part of the fundamentals of what would become the sciences of weather and earthquake prediction in the later part of the 20th century, easily 30 years ahead of scientists. He predicted the existence of a ninth Planet (Pluto) many years in advance, and the discovery of huge quantities of gold on the bottom of the ocean, which was discovered in the 1970's.

He also attributed the main cause of motion (earthquakes) in the Earth's tectonic plates and crust to the motions and relationships of cosmic influences. The trilogy of the "*Return of the Phoenix*" documents how this important clue lead to the author's discovery "*Vortex Tectonics*", which explains how the influence of the orbital relationships and motions in the Earth, Moon, and Sun determine the occurrence of earthquakes, volcanism, and the El Nino weather pattern.

In the late 1920's and early 1930's, Cayce was the first to describe the concept of the shifting of the pole as a result of the crust of the Earth moving independently from the core of the Earth to bring a different surface area over the spin axis. During the past 30 years, this concept has received more and more attention by geophysicists, some of whom now seriously argue that the crust does move independently of the core of the Earth, gliding and jiggling at times over a liquid molten mantle to produce earthquakes and other geological changes. Some geophysicists now also argue that the best way to explain the large variation in sea-levels throughout ancient times and other data is to visualize that the earth's crust moves and shifts fairly frequently and quite suddenly, much more rapidly than previously imagined.

Cayce was equally as far ahead of the archeologists and paleontologists. He correctly described the location of an Essene community which was discovered after his death when the Dead Sea Scrolls were found. He even described it correctly as a community of both men and women, a fact which was not proven until long after his death.

It took nearly another 60 years for the predominance of evidence from around the world to prove Cayce probably correct in his claim that a very old precursor civilization could travel across the oceans and thus influenced the rise of cultures in both the Americas and Eurasia. Though the specifics are still hotly debated, a great many modern researchers and scientists now agree that there is an abundance of evidence in all continents to prove this point incontrovertible.

Closely related to his ability to read science well in advance of its practitioners, Cayce astutely predicted the broad stream of the technological future. Well before their time, before the demands of global war which gave impetus to them, Cayce pinpointed the industries which would expand and eventually become dominant in the post-war world. He had a remarkable eye for the growth sectors of the age.

He was able to fully describe the major technology industries which would drive the development of the American economy after World War II. Consistently, 10 to 30 years in advance of the economy, while ideas were still in their novelty stage, Cayce advised his clients to focus on all aspects of radio, telegraphy, telephones, TV, communications and electronics industries which were associated with war products, deluxe fine furniture cabinetry for electronic products, composite woods and molded materials, plastic resins, and, not least, aviation. David Kahn, one of Cayce's best friends, made millions of dollars with this information and he shared some of it with Cayce and the foundation which inherited Cayce's life works.

As early as 1930, Cayce predicted the long term growth of the aviation industry and the long term expansion of mass communications through radio broadcasting. In early 1932, Cayce predicted the integration of telegraphy and radio communication by 1939 and the switch to using radio (microwave towers) to transmit wireless telegraphs, all of which occurred in various experiments by that date. In March 1944, Cayce predicted that consumer products, especially radio, refrigerators, automobiles, lights, and the like would drive the post-war economy. At the same time, well before a TV broadcasting industry was created, he predicted that TV would become an important consumer product.

Cayce was probably the first human being to describe the concept of infra-red vision. Infrared night vision is now one of the mainstays of U.S. military superiority. In 1933, Cayce described the airplane manufacturing and transportation industries as the most ***"outstanding"*** change which would come, which, in physical terms in North America, is exactly right.

Scientific Accuracy of Cayce's Clairvoyance

For the bottom line, when all is said and done, what is Cayce's rate of accuracy? All of Cayce's objective historical predictions which could be verified in historical records were rigorously analyzed in Books Two and Three of the trilogy of the *"Return of the Phoenix"*. The economic and political predictions were presented as an historical anthology in *"Book Two: The Great Break-Up"*.

Table 200: Accuracy of Edgar Cayce's Long-Term Precognition and Clairvoyance

Based on verification of 253 (N) objective historical predictions and descriptions;
N = net predictions F=failed P= fulfilled PEN=pending
Compiled & copyright 1999 by MW Mandeville
For the key to the columns and notations in Table 200, look for *"Notations"* at the end of this section.

CH.	TYPE OF PREDICTION	TOT	R	U	A	N	F	P	PEN	SCORE
	Long Term Clairvoyance:									
22	Long General Stock Market	48	6	4	9	29	1	28	0	96.55%
23	Total Economic	90	12	3	12	63	4	58	1	93.55%
M	Total World Political	199	21	7	90	81	6	72	3	92.31%
M	Total Earth Changes	167	33	32	35	67	6	32	29	84.21%
42	Total Millennium	53	10	0	30	13	0	1	12	100.00%
	CAYCE'S HARD SCORE	557	82	46	176	253	17	191	45	91.83%

Table 200, from *"The Great Break-Up"*, summarizes the analysis of 557 of Cayce's objective non-personal statements. 82 of these <R> were dropped because they were essentially duplicates, 46 <U> were dropped because there was no historical record which could be found to fail or pass the prediction, and 176 <A> were discarded because they were more appropriately considered as advice than as a clairvoyant prediction.

In those which were left, <N> Net, Cayce gave 253 world-class objective predictions, many of them truly long range prophecy of the chief trends and outcomes of the 20th century. These were predictions made over a period of 22 years and were neither advice nor outcomes which could be guessed or inferred by a knowledgeable observer. At least 191 of these predictions were objectively fulfilled, see <P>ass, and 17 <F>ailed. Still pending are 45 "destiny" prophecies, <PEN>, which are even more important today than when Cayce gave them. Almost all of these pending prophecies are due to be fulfilled during the "millennial period", which began in 1998 and continues on for at least the first few decades of Century 21 (or even longer).

For those which passed or failed, Cayce's rate of accuracy was 91.83%. No other clairvoyant in human history has been scientifically rated with this type of hard score. In Table 200 above, the Column labeled "CH" indicates the chapter number in the *"Return of the Phoenix"* where a summary analysis for that type of prediction is presented as a matter of scientific record.

Most of the economic predictions were for long range trends during Cayce's lifetime, some for a few months, many for a year to a few years. A few were for periods which ranged between several years to several decades. Out of 63 general economic predictions which were specific enough to validate, Cayce made 4 errors. One was about the gross revenues of Western Union in 1933, which fell far short of his optimistic projection. Another was about a sector in the economy which did not do as well as Cayce predicted during one year. The third was his failure to predict the recessionary downturn of 1938. A fourth was his failure to correctly describe the supply of wheat. Four failures out of 63 leaves Cayce with an accuracy rate of 93.55%. for long term economic predictions.

Section Three

The Global Political Shift

Standing on the Southern high point of Second Mesa, The Old Hopi stared way off into the distance, looking the full distance of the 80 miles towards the San Francisco Peaks. He sighed and squinted, hard, as if trying to make out something among the scudding clouds which surrounded the peaks. Twice more he sighed as he turned from left to right and then back again as if to take in the whole tableau of the distant horizon. Finally he continued to recount what the clans whispered only from ear to ear. "Those who take from the Mother and from her children without returning respect to the Earth", he began, "have begun again to do what they did to the peoples of the Medicine Wheels in the past generations. But this time they send their soldiers to the East, not the West, this time they do not look for gold or land to farm, they have sent their soldiers to control the lands and the peoples of the black gold. But the Great Spirit shows all who look that it will not go well for them if they remain there. We, the peoples of the Medicine Wheels, were on this land only a small stream of people. We were easily forded and damned and diverted by the armies of whites who came to plunder the land for the sake of their own desires. But the people of the black gold are many, they are like a huge sea and they cannot be diverted. If the soldiers do not return from those lands, the glittering black gold of greed will poison the minds of all those who do not seek harmony with the Mother. A great many may drown in that sea."

The Old Hopi, 2003
as quoted in "Transmutations", MW Mandeville, 2003

Introduction

Though Edgar Cayce died in 1945, his spiritual work has endured and from this we have gained a great many things which provide important clues for our own lives. One of these clues is about a major historical economic "crunch", a deep recession or even depression, which is due in 2006 or 2007. Based on rigorous analysis of Cayce's life work and several hundred of his clairvoyant predictions about the future, we have the certain knowledge that this "tip" has a 93% probability of being right on the money.

Accordingly, we can use this as a given: we collectively hit the economic "Wall" in 2006 or 2007. The question then becomes, how bad is this "Wall". How hard will we hit it? How can we best prepare to avoid the losses while positioning ourselves to gain for a new era of growth after the Wall? And, not least, what is our best strategy between NOW and then? When will the recession of 2001-2003 lift, where will the new jobs appear between now and then? Can we count on a new wave of growth before 2006?

To answer these questions we need to ground ourselves thoroughly by watching and understanding the shifting international political dynamics which are now shaping the world. We cannot avoid keeping a firm, highly skeptical eye on the changes and trends in the political arena. Sudden policy shifts, tax policies, invasions, terrorist moves, all can, as we have all seen this past several years, profoundly affect the economy, the bubbles, and the running of both the Bears and the Bulls.

Our hyper-reactivity is likely to persist, indeed how could it not? The recessionary and deflationary pressures, the unstable occupation in Iraq, and the onset of the third wave of Al Qaeda's terror bombing, these and a lot of other major conditions demand major actions during the Spring of 2003. Our piles of dirty laundry are huge. Doubtless we are collectively in a period of heightened activity, reactivity, and expectations during which politicians are not only expected to move rapidly to fix things, they are actually aggressively making major moves. The moves they are making will have major impacts on current trends, some negative, some positive. These impacts will shape how economic conditions will work out in 2006 and 2007.

As the international changes unfold, the North American and world economies will be affected, either stimulated to expand or left limp and loosing ground. Or, some may be stimulated while others hang virtually moribund. We need to understand what is driving these changes in order to anticipate the ups and the downs which are coming our way.

By watching the changes carefully, we should be able to catch the drift and adjust our strategies appropriately.

Some of the key trends we must watch are those related to American international actions since 911. America's international political moves since the summer of 2002 have become an increasing problem. Statements, policies, and acts by the U.S. government since then have set in motion many fundamental, radicalizing changes in attitudes and understanding which are now unfolding in every nation more rapidly than ever before. This attitudinal shift hardened with Bush's derisive dismissal of the United Nations over the question of disarming Iraq. Since then, people in large numbers around the world are changing how they perceive Americans and what they think of them.

How much more polarization can the international take? As of May 2003, both the Cold War Era and the Dot-Com Bubble Era are ancient history. Even the "911 War On Terrorism Era" began to recede rapidly into history after the aggressive U.S. invasion of Iraq. Now we are shifting rapidly into another era of awareness and concern as a direct reaction to the precipitous invasion of Iraq. Some of these reactions in Europe and the U.S. are highly opposed to U.S. actions.

These changes in attitudes are profoundly altering international alliances, domestic coalitions, the operations of the United Nations, political activities, and how people do business, work, make money, and think about other nations. These will inevitably lead to changes in how people vote for their leaders and react to the steep economic shifts which are on the way. In many ways, we will all take the "rap" for George Bush Junior.

The shifts in consciousness and awareness which are underway during May 2003 are as yet largely unheralded by the Mass Broadcast Media. Oblivious to widespread "inner" shifts in consciousness and attitudes, much of America's editorial class in the big commercial news rooms continue to serve up a sort of half-baked, faux-centrism which is centered on the yuppie boomer ethos of America's parochial, me-first, "greed is good" corporatized set. Most of the boomer ethos is full of illusions which are dying at an inceasingly rapid rate.

Don't wait to be surprised by the shift in consciousness which is underway. Lead it! Thus, this Section and the next one. The following pages summarize some of the main shifts which currently are shaping political and economic factors in the United States and the World. We begin with the War On Terrorism.

The War On Terrorism

In May 2001, the Dot-Com Bubble in the stock market finally broke after the DJI average reached a momentary high slightly above 11,000. Shortly thereafter stocks lost about 10% in value on the average and American political talk began to center on what to do about a mild recession which had ended the long boom of the Clinton years. But before economic stimulus plans began to take real shape, Al Qaeda's attack destroyed the Twin Towers on 9-11-2001.

The economic reaction was severe and as immediate as the emotional trauma. The stock market dove another 2000 points in value, the airline and tourism industries virtually collapsed, and the economy began to gallup into a deepening recession. As Bush exclaimed many times afterwards, 911 changed everything.

And so he should so exclaim. The attack on the Twin Towers was a decisive declaration of war by Osama Bin Laden on behalf of his Al Qaeda movement. It was his opening demonstration of a master strategy to degrade the American economy and goad the U.S. into over-reactions and increasing military involvement in the Middle East.

In strategy papers which are now several years old, Osama Bin Laden outlined how he intended to use increased American military activity in the Middle East to polarize the Arab world against Israel and the U.S. He believed the U.S. could be counted on to move clumsily, arrogantly, and in ways which are culturally insulting to Islamic and Arabic cultures. By fanning the seeds of dissension which American ignorance and insensitivity were sure to create, Osama Bin Laden postulated that Islamic fundamentalists could unify more Arabs and seize more power in more Islamic countries. In this way he could prepare the ground eventually for an even larger conflict and continue to escalate the influence and power of his fundamentalist brethren.

Osama Bin Laden expected nothing less than to win his Jihad against the Western nations by using a political strategy to out-dance and out-flank the military prowess of the Americans in a completely new form of world political conflict and struggle. He made it plain that he eventually hoped to turn even Saudia Arabia into a despotic theocracy which would be conservatively aligned with a block of newly converted fundamentalist Islamic states against the U.S. He hoped to garner enough power to eventually dictate and impose an "Arab" peace on Palestine as well as eradicate Western cultural influences in the Middle East.

To begin the Jihad, Osama Bin Laden needed to set "an engagement trap" with enough stimulating lures to cause large reactions among the Americans. He baited his trap with the 911 strike against the Twin Towers and succeeded beyond his wildest imagination.

George W. Bush and his Republican cohorts took the bait probably far more aggressively than Osama Bin Laden expected. The Bush administration promised to destroy Al Qaeda sanctuaries in Afghanistan while capturing and bringing to international justice the Al Qaeda operatives who had fanned out around the world to create a terrorist network. Bowing as well to strong Arabic pressure, Bush promised to end one of the main "growth factors" of international terrorism: the Palestinian conflict. He pledged to create an international peace process by the Fall of 2002 which would begin to oversee the installation of a viable state for the Palestinian people.

In the ensuing months, from October 2001 through March 2002, the Bush administration successfully forged an unparalleled coalition of support for a world wide "War On Terrorism". Floating on 50 years of goodwill for American efforts to defend Europe and many other nations from war, most of the nations of the world signed up to cooperate in unparalleled police coordination to root out terrorist cells, seize their financial assets and links, and destroy their training activities and materials.

While creating the worldwide police dragnet, the Bush administration also secured tacit U.N. acceptance of police action in Afghanistan to end the terrible Taliban tyranny and the refuge they had provided for the Al Qaeda terrorists. Riding a wave of universal support and agreement, with no opposition apparent anywhere, the Bush administration plunged the U.S. into a forceful but parsimonious campaign in Afghanistan within 60 days of the 911 attacks.

After about six months of operations, the U.S. military was able to destroy most of Al Qaeda's Sanctuaries and provide enough support to local Afghani forces to enable them to overcome the Taliban tyranny. By the late Spring of 2002, a new government was beginning to form in Afghanistan and the U.S. had begun to take up a campaign to create a final, peaceful solution to Palestine.

So far so good. With almost no domestic political opposition, with widespread international agreement, and with unparalleled high popularity and prestige in all quarters of the Earth (except of course for the so-called "Axis Of Evil" countries of Iraq, Syria, Iran, and North Korea), the Bush administration plotted out its next moves on the international stage.

The Emergence Of The Imperial Faction

In March 2002 the world was still primarily focused on the hunt for Osama Bin Laden and on Palestine, which had become a quagmire of extreme violence on both sides of the fences, for both Jews and Arabs. To diffuse terrorism at all levels, Bush promised to immediately begin to work with Arab leaders to forge a final peace treaty for an independent Palestinian State. After proposing that an international conference would be assembled in the Fall of 2002 to begin the process, he appeared to be gathering up more and more international support in the Muslim world to undertake, finally, serious efforts for the nations of the world to define and insist on "the peace solution" for Palestine.

But then Bush began to shift his focus of attention. He started talking about his tax reduction proposals to re-stimulate the American economy while spending more and more time at his Texas Ranch. And then, by late Spring, for reasons unknown to the public and the media, political conversations by Bush, his officials, and many people in Washington DC began to talk about the need to "get Saddam" and destroy his purported weapons of mass destruction.

Soon the conversation took on the form of a campaign to gain support for a military strike against Saddam, dutifully attended by "Crisis In Iraq" signboards and dramatic musical flourishes on the Mass Broadcast Media. The Media became quite efficient at using these to "create" an event, which more often than not were little more than a "soundbite quote" taken from a staged "fake event". Bush and Company soon could be seen showing up at all kinds of national meetings of business and professional groups where they never failed to get a "soundbite" broadcast to the world about the need to remove Saddam from power.

The "war campaign" on TV soon eclipsed both the military campaign in Afghanistan and Bush's promises to Arab leaders about Palestine. Perhaps the early months of the War On Terrorism were too successful and engendered too much arrogance to keep the policy makers focused on what was already a very large agenda.

As the agit-prop (propaganda designed to stimulate mass reactions) began to flow into the mass mind, a new group in American politics became increasingly apparent in Washington D.C. and New York. During the Summer of 2002, a relatively large number of high-ranking government officials, "think tank consultants", "technical experts", and "media columnists", often calling themselves neo-conservatives, began to publicly wage an intense propaganda campaign in Washington DC to force Saddam Hussein to disarm and destroy his so-called "weapons of mass destruction".

Working week after week through hearings, public speeches, newspaper essays, and a never-ending series of appearances on the American Mass Broadcast Media, the Saddam issue was pushed higher and higher up on the American agenda. Throughout the Summer of 2002, even while ignoring his freshly made solemn promises to create a peace process for the Palestinians, Bush, along with his primary cabinet members and aides, undertook to convince the United Nations and the American public to support American military actions to force the disarmament of Iraq.

Although the job in Afghanistan was unfinished, the U.S. economy was still sinking, most people in Palestine were living from charity day-to-day while both Israeli's and Palestinians lived amidst constant axiety over violence, and Bush's administrators still had not dispersed U.S. funds to fulfill 911 emergency measures and Homeland Defense Programs, it soon enough became obvious that major figures in the Bush administration were caught up in a propaganda barrage about disarming Iraq and the Axis of Evil. Suddenly it was nearly impossible to find anyone talking about the issues at hand while it seemed nearly everyone in government began to talk unofficially about what kind of a military campaign would need to invade Iraq to overthrow the sadistic regime of Saddam Hussein.

By September of 2002 many new internet-media sources as well as many print media publications of various forms were discussing in alarmed tones the emergence of this powerful faction of neo-conservatives who were pushing aggressively for an invasion of Iraq to topple Saddam Hussein. This idea of a possible invasion of Iraq was a lot more than an idea, many new sources warned, it was an actual plan by an organized faction to manipulate the U.S. into an invasion as soon as possible.

The sardonic left described this emerging faction of old cold warriors and new-wave right wing ideologues as "chicken hawks". The chicken hawks described themselves more discretely as neo-conservatives, neo-cons for short. There is something in the word associational matrix for "neo-con" which seems to resonate, but that is another story for another day.

The chicken hawks consistently seemed more interested in using American military forces to dominate international events than in creating solutions for peace. Totally ignoring the bloody brutalities in Israel, a burgeoning, never-ending supply of TV talking-head experts, the type of never-heard-of-before authorities which CNN and FOX and other networks love to dredge up from nowhere to opine "authoritatively" on the topic of the day, a never-ending persuasion campaign began 24/7 to saturate American Mass Media on the importance of removing Saddam Hussein's weapons of mass destruction from Iraq. Seize them before they could be used, was the mantra, let us go into Iraq before Christmas and solve the problem of Saddam Hussein once and for all!

Just after Labor Day, Congress reconvened from its summer break. The talk of war had become so loud, with suicide bombers and Israeli military reactions in Palestine delivering a constant supply of new brutalities virtually every week, both the Republican and Democratic parties soon became embroiled in the conversation about invading Iraq. Since Saddam is so brutal and such an illegitimate tyrant, argued a great many members of Congress, why not present the evidence of these "weapons of mass destruction" and ask the United Nations to undertake to find them and remove them?

But no one could provide the evidence. Even the CIA during September 2002 had to report that there was no tangible indication that "weapons of mass destruction" still remained in Iraq. Nor could the CIA, trying as best it could, honestly provide any evidence of mutual linkage and cooperation between Saddam and Al Qaeda in actually conducting terrorist attacks. Nor could any proof be found about Saddam's cooperation with any active terrorist network. The most which could be said, which was offered up publicly by Saddam himself, was that he was sending $25,000 stipends to help the families of suicide bombers.

That no doubt is hardball, cut-throat politics, but is it terrorism? Is it supportive of terrorist activity? Doubtless such conniving is psychologically supportive of maintaining a culture of permissiveness for terrorism. But should you stop it by undertaking a war which will kill at least several thousand more people?

With no real current evidence available to prove a real case worthy of war against Saddam, and with the probable death from war of tens of thousands of people whose main crime was being in a place where they were oppressed by a vicious dictator, anti-war letters poured into Washington D.C. Soon enough a large number of Representatives and Senators made it plain that they were in no mood to declare a preemptive war to invade Iraq. Instead they authorized the President to undertake action which was appropriate to protect the U.S. from terrorism and enforce U.N. resolutions concerning the disarmament of Iraq.

Congressional Democrats hoped they had derailed the chicken hawks and had forced Bush to work within the consensus of the U.N.

Most of the important members of the United Nations Security Council had already made it clear that arms inspections in Iraq, with more vigor and power to go anywhere in Iraq, should be given one more chance to work. Accordingly, most people reasonably expected that U.N. inspections efforts would defuse war talk and set the world on a course of inspections of suspected sites and negotiated settlements of any remaining armaments issues with Iraq.

By the time the U.N. met to set up the new inspections programs, many new internet-media information sources had begun to report that war was virtually inevitable because it was being instigated by on organized cabal. From internet sources, it was easy to see that the chicken hawks were not a random group of specialists and politicians, they had their own lobby in the form of a real cabal! They had organized around the "*Project For The New American Century*", a group which was constituted as a non-profit "think tank" in 1997. This "PNAC" had developed a set of briefings and position papers which promoted a grandiose vision of how America should undertake to maintain itself as the world's sole super-power. It pulled no punches about the need of the U.S. to dominate the world by force of arms.

This group consciously and boldly advocated taking up the mantle of imposing a "Pax Americana" (their actual words) on the world. They quite baldly referred to the Pax Romana and the Pax Britannia, which were the two most successful and far-reaching imperial epochs of the past. Here is what they said in the imperial dossier (page 9) titled "*REBUILDING AMERICA'S DEFENSES: Strategy, Forces and Resources For a New Century*" which the Faction released in September, 2000: ***"At present the United States faces no global rival. America's grand strategy should aim to preserve and extend this advantageous position as far into the future as possible."*** From earlier Defense Department plans, PNAC obligingly prepared what it described as ***"a blueprint for maintaining U.S. preeminence, precluding the rise of a great power rival, and shaping the international security order in line with American principles and interests"***.

The PNAC group proposed to maintain and extend the American "Pax" with the active use of military might, even as the previous imperial powers had done so. Their proposals detailed at great length how to increase military spending substantially and shift American troops out of Europe and into new regions, especially into the Middle East and Southern Asia (where the most unstable Islamic nations are).

As one looked through the nooks and crannies of the new internet-media information sources, it could be seen, more and more, that the chicken hawks were far more organized than most people were aware. Unbeknownst or at least rarely discussed it at all by the American Mass Media, the new faction had already become the dominant faction in control of the American government. It included in its membership some 25 of the most predominant national security officials inside the internet-media and in front of groups to advocate enthusiastically the various aspects of the tenets and agenda of the long term plan which was outlined by PNAC.

Accordingly, the "*New American Century*" documents rapidly become known and regarded as the basic plan of the Bush Administration to create a dominant, imperial presence in the Middle East. The PNAC group embedded in their briefing documents a large number of euphemistic, progressive-sounding phrases to appear to be talking about enhancing American values. The euphemisms give the appearance of a polite group of academics and consultants who are working for a more peaceful world and a modest increase in defense spending to modernize military technology.

From a casual read of the documents, one could easily conclude that this document is written in idealistic rhetoric and it being used by war contractors to "sell" the government on spending more money on defense programs. This is true, but this is not nearly the complete story. The PNAC documents chillingly build an argument and a specific plan for America, an America which is alien to most of its citizens and is radically divergent from its traditional policies and some of its most important principles from the past.

The PNAC document calls, most radically, for the U.S. to expand and use its military power actively in whatever PREEMPTIVE wars its officials determine are in the best interests of maintaining America's PRE-EMINENCE. Despite the often somewhat obscure, idealistic, academic rhetoric, the PNAC documents boldly assert that America is the dominant power on Earth and should keep it that way by preemptively preventing the emergence of any challenge to that position. The first proposed targets of the plan are "weapons of mass destruction", such as they claimed Iraq possessed, and the international terrorist networks which might use them.

Surely you played King of the Hill when you were a child? Most people enjoy that game but eventually they grow up and mature past such games. Most people realize eventually that the world requires some willing consensus among a plurality of viewpoints and abilities. That 'give and take" no one really quite rules. As the ancient political philosophers taught thousands of year ago, fear of the King On The Hill does not rule people, it is cooperation and mutual self interest which rule nations. The main political question was always, how to secure that cooperation?

Only tyrants, the ancient Hellenes (Greeks) taught, use fear. This, the great classical political theorists taught, makes the people of tyrannical nations miserable. Their politics are perpetually unstable and their states are doomed to revolutions which seek to kill their tyrants.

The Founding Fathers of the U.S. Constitution observed all this and had virtually memorized the political advice of Plato, Aristotle, and their fellow Hellenes (Greeks). The Founding Fathers endeavored to incorporate the values of cooperation and mutual self-interest into the government they set up. They also tried mightily to make impossible any tyranny by small cliques of people. With exacting precision, they wrote into their constitution procedures which made it impossible for the President to undertake aggressive military operations in foreign lands, unless he had first secured a declaration of war from both houses of Congress.

The PNAC documents stray far from these principles and values which are embedded in the U.S. Constitution. The principles which were celebrated in American politics for the greater part of its history are simply ignored, as if they never existed. PNAC's documents constitute nothing less than a unilateral declaration of a World Order imposed by the World's only superpower, the latest successor, one might suppose, to the Roman and British Empires.

The PNAC documents were too discrete to use "imperial" terminology, but here is how the new Imperial Order would work. This resurgent imperial world order would be subordinate to the American government, not the United Nations. The American Presidency would act more or less as the chief Imperial Proconsul of the Empire. Under the doctrines of "Preeminence" and "Preemption", the American Proconsul would strike anywhere at any time to deal with threats to its self-conceived sense of security.

How seriously should we take this plan? We should regard it as already immensely powerful. After all, currently the world is at least pretending to dance to the tune the PNAC plays. Bush Junior certainly does, he seems to relish being the Grand Marshall of the Parade, which he carries on with great energy.

As well, we most likely ought to regard this group as antithetical to democratic and progressive values and policies, frankly as the "natural enemies" of American democracy. They want to consolidate and manipulate power over all the nations of the world. They have no plans for sharing this power, such as with the United Nations. Accordingly, we might do well to label this group as an "Imperial Faction". These words,

"Imperial Faction", precisely describe what the PNAC partisans are about as they attempt to re-make the American Presidency into an Imperial Proconsul whose most important task is to make sure that no other nation achieves sufficient power to challenge his economic and military power.

Is it surprising that the French, Russians, Dutch, Belgians, and Germans, who have some considerable direct experience with Imperial Pretenders, consider this all rather conceited? It took each of those peoples two to three generations to work imperial pretensions out of their national souls. They have some considerable perspective on such notions.

Is it surprising that Conservatives in the U.S. are extremely upset, floundering as of May 2003 for a sense of what to do? After all, Bush's doctrines constitute nothing less than a Presidential putsch against the U.S. Constitution, which unambiguously makes such doctrines illegal.

The defining moment for PNAC's great work came in October 2002. At that time, Congress delegated the power to the President to undertake any action which is appropriate for fighting terrorism and eliminating weapons of mass destruction which "threaten" Americans. With this resolution the Imperial Faction obtained nearly complete control of the U.S. government, including the legally defined right to act dictatorially at any moment in any location for any reason which it, in its sole judgement, acting through the office of the "Imperial Proconsul", George Bush, determined to be appropriate to "protect" America from terrorism.

With this absurd delegation of authority, which is without doubt an unconstitutional grant of power, the U.S. Constitution became moot and the government has thus been highjacked right out from underneath the inadequate mentalities and slothful habits of many of those who constitute the political class in the hallways under the Capital Dome in Washington D.C. This left practically only Senator Byrd and Representative Kucinich, with some assistance by Kennedy, Feinstein, and a handful of other Democrats with Republican Ron Paul, as the only remaining constitutional patriots left under the Dome fighting valiently to keep the Constitution alive.

A decisive victory over Al Qaeda was also highjacked right out from underneath the eyes of the U.S. Military during the Spring, Summer, and Fall months. Al Qaeda groups made several "breakouts" across the borders of Afghanistan while the Imperial Faction was busy gathering momentum for a war on Iraq. Observers in Asia believe that several thousand Al Qaeda partisans (some of them trained terrorists), along with Osama Bin Laden, were able to escape and flee to dozens of other countries because the U.S. had not committed enough troops of its own, nor requested enough support from others, to effectively seal the borders of Afghanistan to capture them.

Thus by October 2002, real things on the ground were not looking nearly as promising as in the Spring. Situations which should have been easily resolved began to have the odor of half-measures made by disorganized, half-distracted people who didn't entirely know what they were doing. Concerning this, the Imperial Faction had not a word except this one: Saddam.

Polarization Of The United Nations

Mandated by Congress to work through the United Nations to disarm Iraq of "weapons of mass destruction", Bush and his cohorts began a discordant struggle with the principal nations of the world. An intense Keystone Cops diplomacy culminated in February 2003 nearly five months later with almost complete international estrangement of the government of the United States.

The Imperial Faction, after a month or so of intense international wrangling about the wording choices for a draft U.N. Security Council resolution about Iraq, finally agreed in November 2002 to the unmovable demands of the French, Germans, Russians, and Chinese. The U.S. reluctantly voted in favor of allowing new arms inspections and disclosure programs to be coordinated under Hans Blix.

In essence, the nations of the world, including Iraq, agreed to pick up where the arms inspections program had left off in 1998. They also agreed to provide enough time to allow Hans Blix and his team to investigate Iraq's disclosures and make decisive actions to verify them and find violations, if any existed. The world authorized Hans Blix and Iraq a period of 90 days to conduct their initial inspections program and make a series of reports about all findings and actions related to Iraq's compliance, leading up to a final report at the end of February 2003 at which time the effectiveness of the program could be assessed..

With widespread material support from the European powers, Blix and his team of inspectors undertook the inspections program with great speed, dedication, and thoroughness. Events seemed to flow positively in Iraq almost immediately with the unflappable Hans Blix seemingly in charge. Reversing nearly a decade of non-cooperation, Iraq finally began to yield to international demands and Hans Blix began to make progress in verifying the facts in Iraq. Quickly enough, the inspectors were able to verify that there was no program in Iraq for making the much-dreaded nuclear bombs.

They repeatedly pushed the Iraqis into complying with the U.N. demands. Soon, expansions of the searches were under way, nothing was "off-limits", surveys by aerial over-flight became intrusive, and individual questioning of Iraqi scientists began to break the deep ice. Progressively, Blix and his associates greatly expanded the scope of cooperation which the Iraqi government gave the international community. They even secured the right to visit Saddam's infamous "Presidential Palaces". Eventually, Blix even persuaded the Iraqis to begin destroying some missiles which could technically fly slightly further than allowed by United Nations guidelines.

Despite the continuing unrest and sporadic violence in Palestine, the world felt upbeat during November and December 2002 about the prospects for a peaceful solution and people began to look for new U.S. economic programs to stimulate resurgence and expansion in the sluggish U.S. and international economies.

But during December 2002, events turned very strange. Bush and his cohorts in the Imperial Faction began a nearly day by day carping critique of the U.N. inspections program and the efforts of Iraq to comply. Every event, finding, or non-finding was nearly certain to evoke critical and sometimes caustic remarks about the deceptions of Iraq, the inability of the arms inspections team to succeed at its task, the inadequacies of the U.N., and bellicose statements about the need of Iraq to surrender immediately what Iraq claimed it did not have.

By January 2003, after some 14 months of great success on the world stage, the U.S. suddenly began to look like a strange and motley collection of Keystone Cops (the name of a 1930's cartoon series of bungling cops who continually trip over themselves and never seem to catch the criminals). The Bush administration attempted to assert, for the benefit of the world, that Saddam's government was "in breach" of its agreements with the U.N. and accordingly, the U.S. and the U.N. should immediately take over Iraq and imprison Saddam. But, despite vast sums of money and control of the world's most sophisticated intelligence networks, Bush, the Imperial Faction, and all the King's Men and all the King's Horses were unable to provide any proof whatsoever that Iraq still had weapons of mass destruction beyond a few trivial technical violations of a few U.N. regulations.

The CIA admitted in September of 2002 that they were unable to provide evidence about terrorism links or forbidden weapons but this disclosure did not stop the Imperial Faction in January and February 2003 from trying to "prove" that not only was Saddam linked with Al

Qaeda and the 911 attacks, he also had enough chemical or biological weapons to pose a "threat" to the U.S. George Bush, Tony Blair, and their cronies in the Imperial Faction attempted up until March of 2003 to give the "appearance of evidence" by submitting a variety of documents, including 10 year old documents, facts which were 15 years out of date, forged invoices, dubious, hard-to-interpret pictures which most international experts dismissed as meaningless, a plagiarized term paper, and a large variety of misleading rhetorical claims, all of which were dutifully reported as "news" by the Mass Hysteria Broadcast Media.

The more the Bush administration attempted to prove its case, the more immature and amateurish Bush and his cronies appeared and the more they were criticized, primarily in the print media, in the new internet-media, and in the international broadcast media in Europe and Asia.

The more the Bush war campaign hit the shoals of increasing opposition through out the world, the more the U.S. and British "Keystone Cops" attempted to float more evidence, seemingly scraped off the bottom of the barrel, to support their demand for U.N. authorization of war against Iraq. The world began to groan more and more in disbelief and discomfort at the seeming political incompetence of George Bush and the strangely naive chutzpah of Colin Powell, the U.S. Secretary of State, in making such insulting presentations.

Finally, some brilliant strategist in the Imperial Faction took aim in the wrong direction and began firing on the United Nations. Soon most of Washington DC was charging that the "old nations" of Europe and United Nations were proving themselves "irrelevant" by not kowtowing to America's "we know best' attitude. The problem was not the lack of evidence, according to the Imperial Faction, the problem was the irrelevance of the United Nations.

Well before hitting the wall of world disbelief, the Imperial Faction proved that it was unable to chew gum and walk at the same time. The U.S. abandoned all of its promises to make peace for the Palestinians. While plotting its war objectives in Iraq, the Imperial Faction gave complete support to Sharon to suppress both Palestinian terrorism and the Palestinian Authority, which Sharon undertook with unparalleled haste and, for an Israeli government, uncharacteristic brutality.

The completely unthinkable became common place and the Palestinian townships turned into ghettos which began to crumble more and more into bombed out shells reminiscent of the unrelenting process of random destruction which destroyed Lebanon during the 1970's and 1980's. Within this routinized daily random destruction, something like

2.8 million people (estimated by BBC News) to 3.5 million (estimated by Sharon's representatives) continued without much hope to live a futile life in a vast ghetto jail, virtually unable to do much of anything except persist blindly and live on the dole.

The U.S. also lost its remaining focus on Afghanistan. In balance, it became reasonably clear in the Fall of 2002 that the U.S. had acted in Afghanistan in an incomplete and incompetent manner. Without sufficient resources, the new government was not able to establish control of the country. Because the U.S. had committed insufficient troops to the task, thousands of Al Qaeda operatives were able to escape into other countries. Osama Bin Laden, who may have been captured or killed at Tora Bora if sufficient troops and planes had been in play, had escaped along with most of his top leadership structure.

The U.S. also began to forget more and more of its promises. By late Fall 2002, Pakistan was still waiting for the money it had been promised as an ally of the U.S. against the Taliban, Warlords were back in virtual control of Afghanistan's rural provinces, the Afghanistan's national government barely could control Kabul and a few major cities, and, despite billions of dollars of satellite spy high technology which can observe such things, the peasants of Afghanistan were readying a bumper crop of poppies to supply the world's heroin markets.

Thus leaving the "old issues" of Palestine, Afghanistan, and Pakistan to their own fates, along with campaign 2000 promises to revitalize American education, the Imperial Faction focused the late Fall 2002 and early months of 2003 almost exclusively on Iraq. Somewhere along the way the Imperial Faction also ignored its first-year-in-office promises to Vincente Fox, the President of Mexico. Rather then take up how to fundamentally restructure U.S.-Mexican border affairs and rights to cross border travel, the conversation was put on hold and they forgot to return to the phone line.

Unbelievably, the Keystone Cops Capers unfolded and broadened to even greater stunning depths as the Imperial Faction strove to forge what it called a "coalition of the willing" to demand immediate U.N. sanction of a U.S. led invasion of Iraq. Early on in this campaign to construct his coalition, Bush formally annunciated his Doctrine of Preemption, under which he claimed the U.S. had the right to invade at will any country which he deemed to be a threat to American lives or interests. Shortly thereafter the Imperial Faction unleashed a cut-throat campaign of extreme manipulation of the leaders of scores of nations. Offering substantial bribes on the one hand, and, on the other hand, not so slyly warning of dire consequences of enmity, loss of aid or trade agreements, the Imperial Faction began to bully nations both large and small to support the U.S. with their U.N. votes, or to provide use of their military bases, or to allow air fly-over rights for U.S. military jets, and so on.

Virtually unparalleled in international diplomacy, the Imperial Faction also began making increasingly bellicose public statements to the nations of the world, demanding with greater and greater arrogance that Europeans and the U.N. Security Council must not wait for the results of the inspections programs. In the face of mounting daily evidence to the contrary, U.S. officials baldly claimed that the inspections process could not possibly work. If effect, the Imperial Faction haughtily insisted that national leaders should ignore the majority opinion of their own peoples (who overwhelmingly wanted no part of an U.S. led invasion of Iraq) to unify the world as soon as possible in support of American intervention in the Middle East.

But the harder the Imperial Faction pressed, the more the world resisted. The world press, and then eventually the U.S. print press, began to ridicule this blatantly sophomoric campaign of mal-adroit deal-making. Soon people around the world were calling it the "coalition of the coerced" or the "coalition of the bribed". The newly-elected democratic government of Turkey, whose Islamic people solidly opposed an invasion of Iraq, rejected Bush's demands and ignored Powell's $26 billion in bribes.

The U.N. Security Council and the Euro peoples unified all right, but it was not in favor of George Bush's War On Iraq. Ironically the Security Council ended up doing EXACTLY what it was set-up to do. Which was? To mediate between states and constrain those who threaten war by finding peaceful solutions from a consensus of the major nations of the world. One might suppose that the arms inspections program provided just such a peaceful program, which the Security Council was quite willing to engage aggressively in Iraq for a long time. The Europeans clearly had in mind conducting the inspections aggressively enough to prevent Iraq from conducting any manufacturing while the inspectors looked for any conceivable ways and means of manufacture.

Finding no credible threats in Iraq, the Security Council and most of the world, including America's nearest neighbors, Canada and Mexico, forthrightly refused Bush's pompous assertion of the "doctrine of preemption". Most of the world recognized from their lessons in History 101 that such doctrines were always (and only) the doctrines of those who would maintain Empire. No doubt both the Canadians AND the Mexicans remembered episodes during earlier centuries when American Jingoes foraged into their territories under the flag of Manifest

Destiny (which amounted more or less to the doctrine of "to take from others, conquer, and be as big as we want because we can").

Under such doctrines, peace is never maintained for long, rather wars proliferated, first for the Proconsuls of Rome, and then for all of the would-be successors among the Spanish, the British, the French, the Russians, the Italians, the Germans, and many others, wars upon wars upon wars to conquer all those who might be a threat, Hail Caesar, Conqueror of the World, Maker of Peace, Sieg Hiel, What A Crock of Crap!

"Et Tu, America?", asked the World-In-Peace. ("You too, America? ") This of course is a paraphrase of the famous Shakespearean line where Caesar rhetorically asks of Brutus, "Et Tu, Brute?", while Brutus, the so-called Republican defender of law, stabs him to death along with a dozen or more members of the ancient Roman Senate. With this collective elmination of what the Senators perceived to be Caesar's threat to the Roman Republic, the peace and relative prosperity of the Roman Republic was split into opposing warring factions. Rather than preserving the Republic in peace by cowardly murdering a leader they feared might force his way upon them, the conniving Senators doomed each other to fight to the death while wreaking carnage across much of the Mediterranean World. They fought for several years before one victor emerged to impose the Roman Empire on the world.)

Unfortunately, the home-grown American Imperial Faction was not interested in the rule of international law to prevent struggles for imperial domination by factions vying for control of the world. They were not at all interested in maintaining the actions of the United States within its treaty obligations under the United Nations Charter, a Charter which the United States had largely underwritten after the close of World War II to provide a standard of international law and a legitimate mechanism for the use of military might of nations against other nation(s).

Though the Charter clearly requires the United States to work within and in accord with the Security Council, the Imperial Faction was more interested in selling Pax Americana. The Imperial Faction was much more interested in making the world acquiesce to U.S. domination in order to change the psychology of the world, simply because it can. Accordingly, they escalated their accusations that the United Nations and even "old" Europe were now irrelevant to the making and keeping of international peace.

The blatant maneuvering during February and March deepened the Keystone Cops Caper into a farcical political tragedy. Empty, rhetorical exercises became more and more common. During the last fumbling, bumbling weeks before the invasion, the Imperial Faction shifted its arguments almost entirely to blatant lies, such as linking Saddam Hussein to the terrorist networks and the attack on the Twin Towers in New York. Week in and week out, the insinuation was constantly made that somehow they were all linked and Saddam bore some personal responsibility for the deaths in New York.

A disturbing percentage of the American public which gets its primary news from the Mass Broadcast Media bought this blatant lie, or perhaps we should call it a deliberately planted hypnotic suggestion. Pollsters say some 50% of the public bought into this lie and were thusly easily manipulated by Mass Media Hypnosis into professing support for George Bush, the Troops, and getting rid of Saddam.

It is of course all linked, IF you accept the primary premise of the Imperial Faction, namely, anybody WE label a terrorist is a terrorist and the only good ones are dead ones.

Unequalled in history, with nearly unlimited power and influence, George W. Bush had a choice to take the high road to making peaceful solutions, beginning in Palestine, or the low road of violence and war to force onto the world his desire to destroy Saddam Hussein and the evil chemical or biological weapons which he ***<might>*** possess. With more power and ability than any man in history to act on the highest moral plane to empower truth, create justice where it had been long denied, and make peace in the world's most violent neighborhoods, George W. Bush choose to make war premised on a thickening, sickening pile of lies and assorted duplicities, even despite his claims to believe in the authority and moral command of the Christ to forgive enemies and transcend violence.

The more the Keystone Cops tragedy unfolded and broadened to demand immediate U.N. sanction of a U.S. invasion of Iraq, the more the world resisted. By February 2003 it was clear even to Bush that the Imperial Faction had severely polarized much of the world against his policies and had mobilized a major portion of the American public in opposition to his proposed War On Iraq. By early March 2003, Bush and his American "Shock and Awe" troops, began to be described in many European quarters as "Stormtroopers", which is the same name which had been given to Germany's World War II Nazi blitzkrieg armies. Bush came to be widely regarded in many areas of the world as the greatest current threat to peace, a worse menace than Saddam Hussein.

This mattered not at all to the Imperial Faction. Intent on reshaping the world's attitudes through the sheer exercise of making the world accept U.S. domination, precisely as articulated in its documents, the Imperial Faction loudly declared the United Nations irrelevant. It proceeded to undertake an invasion of Iraq with a huffed-up "Coalition Of The Willing" composed of U.S. and British troops, a few Australians and Poles, a Danish submarine, and a bizarre list cooked up by PR flacks who claim some 45 countries (and counting) support the U.S. effort in mystical ways.

A huge number of countries watched aghast at this new advent of American Imperialism. By no means mute, except to the ears of the U.S. Mass Broadcast Media which suddenly forgot about the rest of the world, the world began to talk to itself in new ways. The French, more than a small number of Brits, the Germans, and many others made quite clear they regarded Bush's operation as entirely mad. The Russians discretely whispered for Euro ears only about the need to create a countervailing block of power to outweigh the U.S.

A most remarkable transformation of the world scene. One's jaw can only drop in awe at the ability of a single human being, in this case one George W. Bush, to so decisively snatch such a resounding strategic diplomatic defeat from the jaws of the great victory of world support for the "War on Terrorism". The United States has suffered the worst damage to its international reputation since the Vietnam War when it undertook to carpet bomb peasant farmers or to spray their trees and crops with huge quantities of deadly toxic chemicals.

All of the King's Men and all of the King's Horses combined cannot put this Humpty-Dumpty back together again. We are stuck with this damage so long as George W. Bush remains in the Presidency and even his replacement will not restore American credibility very quickly. The massive loss of international prestige and credibility will fundamentally effect the unfolding political and economic dynamics of the next several years, if not for the next generation as a whole.

Henceforth, American initiatives will automatically be seen as self-serving, aggrandizing moves. Every creative act will be seen through highly suspicious eyes. A critical chorus will accompany every move and this will be given major play around the world. It is very difficult to be an effective leader under such circumstances, unless one is willing to rely on using pure bully power in the back alleys.

The economic bottom line now can be put most succinctly. To the extent that the health of the Amerian economy requires international cooperation, we have a big problem To deal with this problem, the international political dynamics which have been set forth in this Section must be understood.

Polarization of Eurasia and the Middle East

During the early weeks of 2003, it seemed as if everything was holding in place, waiting in a state of chaotic flux for something, anything, finally to happen. It seemed that anything could happen, even a rational consensus to allow the United Nations to work without war. But finally in April 2003, with the invasion of Iraq by American and British troops in open insolence of the United Nations, some of the confusion of the previous six months finally began to lift. As the occupation of Iraq began, some of the most probable dynamics for the next several years could finally be surmised.

Over-riding everything for the next several years will be the strong polarization and fragmentation of opinion which the invasion of Iraq has created throughout the world. The strong demands by the Bush Administration for an immediate solution to the "problem" of Saddam Hussein has made the U.S. appear to actually desire conducting war and this has aroused a strong and abiding suspicion about U.S. motives which is not going to go anywhere anytime soon.

This "appearance of war fever" was made much worse by the vigorous support which the U.S. Broadcast Mass Media gave to the Bush Administration's invasion efforts. As a result of hyper-ventilated Mass Broadcast promotion, which included broadcasting many months of questionable claims and untenable proofs as "news", the world perceived the strong push by the Bush Administration to invade Iraq as an unseemly strong "desire" to wage war. Bush's disclaimers about being "a patient man" , or as not having made the decision until February 2003, seemed to be merely another political hypocrisy by a man who is now seen more and more as a compulsive manipulator who is more than a bit out of his intellectual league and can no longer keep track of his lies.

Since ancient times, those who have been perceived to "desire" war have been the most suspected and the most opposed. This problem is now fundamental to Bush's image, indeed to the entire Imperial Faction. This "war image" will shape all reactions to Bush's policies and invoke deep suspicion and fears.

Fear and jealousy drives some of the opposition but there is far more to the opposition than just emotions. Pretty much, it boils down to elementary common sense. Who knows who will be next on the aggressor's list? No ideology is necessary to answer the question and the answer brings dissimilar nations to band together to oppose wars of aggression.

Banding together to stop war is easy for people of all nationalities and orientations to rationalize. Most organized religions tend to lead people to oppose arguments which favor war. Nearly all systems of morality reject the legitimacy of desires for war. War as such is seen as illegitimate, a terrible evil. Thus, invariably, religious leaders appeal to higher principles of being and universally proclaim that wars of aggression must be resisted.

Accordingly, on the basis of emotions, sound logic, and ethical principles, nations since the earliest of times have been quick to band together in opposition to the aggressive war efforts of any nation. Was the strong opposition, which included the primary leaders of most religions, to the invasion of Iraq during the last half of 2002 and early 2003 all that surprising? Wasn't it inevitable that the Bush Administration's aggressive invasion plan should be received with extreme caution and finally with strong negativity in many quarters of the Earth?

A simple "defensive" reaction to an aggressive action by Iraq, such as the repulsion of Iraq from Kuwait, would have met with widescale, instant approval. But the sudden "desire", some ten years later, for an immediate, aggressive invasion, along with sophomoric, clumsy bullying and bribing tactics by the functionaries of the Bush Administration, was received with great distaste by both the governments and the populations of several billion people, as well as by about two thirds of the American population.

The personal style of the Bush Administration compounds this polarization and makes the reactions even worse than they need to be. In dealing with gravely complex issues, the rhetoric is almost always cast in simplistic, absolutist terms: "you either for us or you are against us". The absolutist terms are often pitched with hardball, elitist attitudes: "if WE think you are supporting terrorists in any way, we reserve the right to take you down in any way WE decide". Its officials appear to offer only a haughty, take-it-or-leave-it proposition.

All this and much more has been seen and taken as bullying by many countries and world leaders during the past few months. Can it be any wonder, then, that North Korea has announced to the world that small countries like itself must have nuclear weapons to defend itself from the world's biggest bully? Given the behaviors and rhetoric of the Imperial Faction, how could they not conclude that North Korea needs the big bomb to protect itself?

In addition to the use of mean-spirited coercion techniques, which most of the Imperial Faction appear to have sophomorically mistaken for negotiation, George Bush often paints a patina of being selected for a personal destiny, of personally receiving God's guidance to protect the American people and their unique role as the leader of the world. Even if it were true, a good portion of the world tends to choke on what it regards as the vanities of a spiritual simpleton. Who can argue with the fervent prayers of the Pope for God to intervene with peace to stop the invasion of Iraq? The personal divinely-inspired destiny of George W. Bush? How can anyone be so stupid as to put themselves into that situation? All such thoughts about God's leadership are better left for private conversations with God, where they might be properly handled.

The problem of Bush's public image gets much worse than being perceived as a hardball player with a penchant for tall claims and inappropriate use of religion in support of war. Whether it is true or not, ex-alkie, ex-cokehead George Bush now appears to have a major part of his personal identity invested in the need to exclaim publicly his special relationship with the Divine Right Of Manifest Destiny, Version 4.0.

Accordingly, talk of more divine destiny is probably coming down the pike. Indeed, Bush told Palestinian Prime Minister Mahmoud Abbas during the June 2003 summit conference in Aqaba that "*God told me to strike at al Qaida and I struck them, and then he instructed me to strike at Saddam, which I did, and now I am determined to solve the problem in the Middle East. If you help me I will act, and if not, the elections will come and I will have to focus on them*". (As quoted in a news report by Arnon Regular titled "Road map is a life saver for us, PM Abbas tells Hamas", which appeared in Harratz.com (an Israeli newspaper).

As will be seen in the next section in the discussion of Media Sorcery and American Jingoism, this portends more difficult moments with Imperial Brinkmanship. On the one hand, some will continue to clack behind his back about what a mean-spirited wacky nut Bush is. On the other hand, many in the Islamic world will be continually reinforced in their belief that Bush really is another cruel "Crusader" seeking to kill innocent Muslims and rob them of Islam while turning their land into a Satanic servant of Godless materialism.

What matters here is not the "truth" of these views. What matters is the "perception" of George Bush. At the moment it is a no-brainer no-win. Unfortunately both for Bush and the United States, Bush has painted himself two undesirable images, either as a wacky nut who is beyond his level of personal competence or as a cruel crusader against Islam. He most likely cannot change either of these images, all subsequent events will be read out within their terms, according to the political proclivities of the perceiver. If you tend to believe in either of these images, will you seriously undertake to take a major political risk in dealing with Bush? Or would it seem preferable to wait him out, hoping that he will not win re-election?

Many (those who have felt unduly pressured or bullied by Bush) may now be tempted to merely wait him out, if they can. Many international leaders may increasingly drag their heels and do everything they can to block Bush's moves, hoping that American public opinion will sour in the lagging economy and shove him out of office during 2004. Already, congressional Democrats have begun a bitter campaign of refusal, delay, and filibusters to stall out Bush's attempt to pack the Judicial system of government with neo-conservative appointees who may be members of the fascist-oriented "Federalist" society. More likely than not, the same sort of opposition will dodge Bush's steps around the world.

From international sources, it appears that another unflattering perception of Bush is growing more prominent. Indirect comments from international sources, perhaps rumors is a more accurate term, suggest that some international leaders regard Bush as a rather naïve, disorganized dilettante with somewhat scattered thinking processes who makes pushy promises but rarely delivers on his side of the proposition.

Some people, including a professional psycho-therapist, are openly speculating that this, along with the hesitant style of choosing words which Bush often displays, or his "vacant" reading of his radio addresses, and many other characteristics, comes from the "disassociation" which comes from the alcohol-damaged brain of a "dry drunk" and/or from the use of Prozac and/or other mood enhancing anti-depressent drugs.

For many others, it appears that Bush has the political style of the one-night stand who never calls again. His elaborate wooing and subsequent dumping of foreign relationships with the leaders of many nations has earned him a large growing "wake" of disillusioned national leaders who no longer regard Bush's comments as "credible". Certainly we know this to be the case with Vincente Fox in Mexico, a good portion of the U.S. education establishment, the leaders of Afghanistan and Pakistan, and the officials of many state and local governments who had to wait for over a year to begin receiving Homeland Defense funds they were promised to beef up metropolitan and state security.

Rumors are, the list is even longer. This "credibility" issue is so widely shared that Turkey's Prime Minister insisted that Powell's $26 billion bribe offer be put in writing before he would submit it to his parliament for ratification.

This polarization has been widened tenfold by the growing awareness of PNAC's manifesto of "*The New American Century*". It was prepared in 2000 but it received little public notice until a few internet-media writers took notice in 2002 that the coming invasion of Iraq had been virtually scripted in the manifesto. George W. Bush's diversion of the War On Terror into an invasion of Iraq comes right off the "plan" pages of this document, right along with Bush's annunciation of the doctrine of "preemptive military strikes against possible threats". Also found on the pages of this document is the advocacy of using American military power to dominate and/or reshape most of the governments of the Middle-East (including Syria and Iran) to realign their policies away from Islamic fundamentalism and terrorism.

Though written in the high-sounding tones of policy-academese, the document is essentially an update of the worst type of Theodore Roosevelt's jingo-thinking (me first, because I can, I have a big stick, watch me whump and take over nations I disapprove). Such thinking produced America's era of overt imperialism in Latin America and Asia during the first half of the 20th century. Wiser, more civilized minds created the United Nations to guide and constrain such tendencies, but it appears fifty years later that America is having yet another struggle with its dark jingo side.

For those who have forgotten, a jingo is a zealous patriot who advocates the sort of "blind patriotism" which advocates "my country right or wrong". Jingo is a British word which is said to have been applied in the late 19th century to describe Disraeli's policies against Russia's Navy, whom he would not allow to use the Mediterranean Sea. That was a rather imperious policy, even to the imperialists.

Since then, those who have been called jingoes are usually those who express or advocate hostility toward other countries, quite as in the manner with which some expressed condescension and/or advocated punitive measures against France for its refusal to allow the U.S. to have U.N. backing for an U.S. invasion of Iraq.

The word was commonly used during the first half of the 20th century to describe American attitudes about the native peoples and the Hispanic nations south of their border. Americans asserted their right to interfere in the affairs of Latin American countries, largely because they saw these peoples and their governments, like the natives before them, or the not quite really free blacks, as inferior and thus undeserving of the same rights and sovereignty as "white men".

PNAC's plan for the "New American Jingoism" is intended as a long term strategy to guide the U.S. government in an active remake of the nations of the Middle East. Given the explicit emphasis on achieving domination by force of might, the new doctrine can easily be seen as basically similar to the Monroe Doctrine, by which early Americans asserted, with not much ethical or moral standing, the right to interfere anywhere in the Americas to keep European powers from gaining an upper hand in the "New World". Under this doctrine, several decades of "gunboat diplomacy" were used to manipulate all of the nations South of Texas.

Many governments of the world are already recognizing the long term implications of having to deal with this new "play" of Manifest Destiny by the Imperial Faction. Since this will be, more of less, the fourth major wave of foreign expansion undertaken by America force of arms, we can call it Manifest Destiny IV. Accordingly, Manifest Destiny IV is now playing worldwide with a loud accompaniment of trumpets sounded by the Mass Broadcast Media and with a chorus of critical catcalls from many portions of the world press.

Unfortunately, the problem of Manifest Destiny IV gets worse. Many prominent intellectual Jewish critics have begun to point out during early 2003 that the membership of the "Project For The American Century" is disproportionately composed of American Jews. Some Jewish analysts are pointing out that PNAC's "vision", especially as espoused by PNAC'S Chairman, William Kristol, seems to be about creating and maintaining an American military policy which serves Israel's foreign policy objectives rather than traditional American roles and principles in the world. From this it is easy to argue that the New Wave Manifest Destiny IV is not American at all, but rather is a form of old-fashioned Zionism. (See Zionism in the notes for this chapter)

In the historical fabric of ideas and peoples, the PNAC Zionism it is not a particularly "Jewish" vision, it is rather more like a sort of secular Israeli fascism with foreskin and bad attitudes about observing things which are truly holy. Perhaps it is a sort of Sharon Likud Zionism, which places the security needs of the State of Israel at the center of the universe.

This of course could explain much of the past six American months (October 2002 – March 2003), including the mystical linkage of Saddam with the international terrorist networks and the new "Axis of Evil" targets of Syria and Iran. The linkage begins with Israel. The linkage ends with Israel. The linkage is Israel. Ooooohmmmm. The Ameican Mass Media will not tell you this. Ooooohmmmm.

A great many organizations and nations in the Arab world are enemies of Israel, ipso facto they are all linked in the Jewish mind, if you have a paranoid one. Never mind that Saddam and Osama hated each other's guts, if they hate Jews, they link. In such a mind it is easy to rationalize converting the "War On Terrorism" to an invasion of Iraq, it is all the same struggle. If you are primarily focused on Israel's security needs, the unpredictable Saddam, the Evil Lunatic of the East who once fired Scud missiles into Israel and who rabidly supported the Palestinian cause with a frothy mouth, was one of the next logical targets of the war on those who terrorize Israel.

But it made very little sense from a North American perspective. Unless.... Unless you could be led to believe that Saddam was partly responsible for the destruction of the Twin Towers.

From a North American point of view, the main problems with this line of reasoning are of course elementary. The main problems include the fact that Saddam Hussein was not involved, never threatened American interests, hated Osama Bin Laden and the other fundamentalist terrorists, never cooperated with their deeds, and, apparently, much as he said, no longer possessed weapons of mass destruction and thus posed no threat at all to American interests.

These are severe problems and make impossible, under solemn international treaties and the United Nations charter, an invasion of Iraq. Unless... Unless, like the infamous Dr. Kissinger (who cannot travel to parts of the world because he is under international indictment for crimes against humanity), you are prepared to lie to God on high while crossing your fingers behind your back.

Accordingly, there was exactly NO reason for Americans to disavow a century of international law and principles to unilaterally undertake an invasion of Iraq in the face of world opinion and the fervent public pleas of the primary leaders of all the great Faiths of the world to "give peace a chance".

The Drama Continues

From all of this, and more, the majority of many of the media and publics in Eurasia are strongly in opposition to what is becoming forthrightly called American Imperialism. Even America's closest neighbors, both the Canadians and the Mexicans, correctly saw this simple reality, that Americans had become obsessed with an illusionary bubble of imperial pretentions spun by narrow cliques.

Americans need to ask why they did not see the fundamental reality in sufficient numbers to prevent George Bush from leaping into Osama Bin Laden's clever trap. As events unfold, they will be given plenty of opportunity to connect the dots. These themes, "American Imperialism" and George Bush's "Imperial Faction", are becoming more and more prevalent in news stories from other countries, throughout the internet news media, in portions of the vigorously spunky print media, and around the world among protest marchers and demonstrations related to global issues.

More probable than not, the Imperial Faction expected all this opposition and criticism to blow over after a make-over of Iraq becomes seen as a success. It is true that dissension about foreign intrigues has often waxed indignantly and then quickly faded away. After all, opposition to the first Gulf War quickly faded away because most people, including this author, supported United Nations international law to control agression against others.

But this Iraqi intrigue is of a different magnitude on a different level in the center of a tinder dry Islamic world which is bitterly opposed to American dominance in the Middle-East. Already it can be seen that the make-over will take many years and so nothing is going to blow over anytime soon. Scarely two months after "occupying" Iraq, it was obvious to the entire world that the U.S. was still fumbling and bumbling to come up with a plan of occupation. So nearly as it can be determined, the only solid part of the plan which was thought out in advance was Haliburton, Vice President Cheney's <previous> company. Haliburton is taking control of Iraq's oil fields and will sell its oil while refurbishing the plant and equipment.

Meanwhile, at the beginning of June 2003, 25 million people in Iraq lived hand-in-mouth without a clue about what was coming next. Saddam was widely regarded as alive and well and busy making audio tape greetings. His faction, along with others, were already re-organizing even as random sniping against American military targets was increasing in frequency, sophistication, and deadliness.

Though no weapons of mass destruction could be found, U.N. arms inspectors were being refused entry into Iraq by the U.S., a nationwide debate was beginning to surface in the U.S. over the NON-evidence of weapons of mass destruction, and the U.S. Military Commander of the invasion, breaking the historical norm (thus sending a profound signal) General Tommy Franks began to "retire" substantially early in the program of taking over Iraq.

The political situation in Iraq up until July 2003 had been so badly handled that some analysts now expect that the U.S. occupation will remain in place for many years, perhaps 10 to 15 years, and that more U.S. troops will die in "liberated" Iraq than during the invasion. This is NOT success. It is confusion engendered by incompetence heaped upon incompetence. Perhaps Frank's army, which is a suburb organization, should also retire from Iraq and be replaced by the officials in the White House and the Senators and Congressmen who authorized this awful state of affairs, along with the Mass Broadcast Media reporters and editors who promoted it.

This sad state of affairs and much more, including the even worse state of affairs in Afghanistan, has inspired a tremendous loss of confidence about the competence of American officals and the U.S. occupation in Iraq. This, along with many other issues of polarization caused by egregious American bullying on the road to the invasion of Iraq, has already created profound, irreversible changes in the world and these will now unfold progressively, more and more unpredictably, in a growing tide these next several years.

Already this little imperial intrigue has generated a huge, record-setting worldwide wave of trans-national, trans -ideological, and trans-religious opposition which is squarely focused on the very real issue of American Imperialism and old-fashioned Jingoism which PNAC has done the world the great favor of making publicly explicit with its manifesto. It is this universal reaction against imperialism which fuels the polarization which now the incompetent occupation in Afghanistan and Iraq continues to deepen and harden.

This polarization of people and attitudes will not disappear anytime soon. It will take 8 to 10 years, at least two Presidential terms under a President who explicitly disavows the doctines of the Imperial Faction, to overcome the damage which the Imperial Faction has created during the past one year.

In the meantime, the highly charged political polarization will cause sudden unpredictable economic moves and shifts on the international scene. Watch Europe and Russia. The Eurasians play a fundamental role on the stage of history and their actions will affect all outcomes. In terms of sheers numbers, they are far more prominent on the stage of history than the Americans. Like in previous generations, the Eurasians could easily stampede in ways which are harmful to North America. If sufficiently frightened by bellicose rhetoric from American mouths which have more ego than brains, Eurasian reactions could thrash economic recovery or growth and they could set in motion long lasting trends which will profoundly affect an industry, a people, a nation.

Most especially, watch every political pronouncement and action which appears to rationalize and advance Manifest Destiny IV, or which diminishes its influence. Observe how it plays in the world. Your life, your prosperity will be fundamentally affected by it.

Section Four

The Global Media Shift

The Old Hopi sighed and stirred the fire to keep it going. "We got a lot more to think about", he said sadly. "We haven't even got to what the words from the sly foxes in the boxes have been saying and doing." He pulled down another well worn book and paged knowingly to a special place. He adjusted his glasses and began to read slowly: "Of course the people don't want war... Naturally, the common people don't want war; neither in Russia nor in England nor in America, nor for that matter in Germany... that is understood... But voice or no voice, the people can always be brought to the bidding of the leaders. That is easy. All you have to do is tell them they are being attacked, and denounce the pacifists for lack of patriotism and exposing the country to danger. It works the same in any country."

Hermann Goering
to Gustave Gilbert, author of "Nuremberg Diary"
Farrar, Straus and Company, 1947, pp. 278-279.)
as quoted by The Old Hopi, 2003
in "Transmutations", MW Mandeville, 2003

Introduction

The profound economic and political changes which are sweeping the globe at many levels are amplified or repressed in countless ways by the global media on a daily basis. To understand the shifts which are sweeping the globe and the likely economic conditions we will have to confront, it is essential to understand how the global media influences and shapes the world. The media does not just report on what is happening, the media shapes and defines what is happening and what can happen.

This is an awesome power and it is not nearly so benign nor objective as we are cajoled into thinking. Some cliques have learned how to practice this power as a dark art of sorcery to conjure up spells of illusion and emotive responses which channel the flow of history. Some of them are politicians, some are terrorists, some work deep in the bowels of the CIA and the Pentagon and similar agencies around the world, and more than a few populate the boardrooms of corporations and the halls of advertising agencies.

For any individual, group, or movement to realistically address the future, these mass media powers and how they have been used must be understood. It is well worth spending the next hour getting the basics of this screwed on right. It is not very difficult and I will use no cheat words, I will pull no punches.

There is no turning back, this Section is terminal for superficial reality systems. When you are through with this essay you may hate the author, for these arguments are designed to ruthlessly strip away from your mind certain cultured veils of illusion, especially the ones which are most comfortable. They are the deadliest of all illusions.

You may end up in pain, but you will at minimum be partially liberated from decades of conditioning in the mind trance of the Republicrat "American Dream" which is re-run perennially by New York's mass media outlets.

We will begin with the simple basics of omission and selection for defining the news without seeming to be anything other than an intrepid reporter. From there we will move to "the slant" by which reporters and editors lead the minds of their audiences in preferred directions. Then we will examine how the arts of mass manipulation have been used to channel the flow of mass reaction, concensus, and historical action through the same perennial "story lines".

Within the American context, the most important story line of all has been the "American Dream" within the perennially reappearing play of "Manifest Destiny". We will look at many of the repeating acts of this play, from the early 19th century through to the Imperial Faction's invasion of Iraq. By seeing how this play is recycled every generation or so with the same ploys, the same tactics of selection and ommission, the same claims which slant and warp the minds of naïve people, and with essentially the same vision by a similar powerful elite of super wealthy people and "special commercial interests", we will come to a much deeper understanding of the true dynamics in the world. We will also be far less susceptible to the dark powers of the "Sorcerers" who pretend to the control of history.

Selection

The first basic of mass manipulation is selecting what is and what is not to be observed. Reporters and editors react to events and amplify their significance by selecting only a few of them to write up and transmit as "stories" to mass audiences. What the media people select, and how they react to what they have selected, profoundly defines the nature of what most people think is going on.

The second basic of mass manipulation, and don't ever forget this cardinal point, is that with its selections (and ommissions), the mass media sets the basic agenda of politics. It does this far more decisively and with much greater impact than the comparatively feeble efforts of politicians or political parties. How and when people respond to WHAT events, trends and economic signals is greatly influenced by what media is available to them, the content which is served up, and, most especially, the attitude or slant with which the information is shaped into "views" of the "news".

The media presents, purportedly, the agendas of politicians and various groups in society. But in the process of selecting the politicians about whom they will report, the media actually shape and set the public agenda. During this process, one of the greatest tattle-tales of their own agenda in "creating" the news is found by observing the "who" and "what agenda" they select to tell about. Which candidates are ignored, or given much less coverage than others? Which candidates get the most coverage? What issues get immediate attention? What issues are never reported on? The answers to these questions do not tell you about the news, definitely not what may be important, nor even what it is that people, the audience, want to know. These answers tell you what the editors and reporters want to put into your life to change your consciousness and behavior.

To see how the process of selection works to define the news without even seeming to, observe how editors and reporters present the "news" related to politicians, political groups, and their controversies. Do editors and reporters select right wing think tanks along the Potomac funded by the ultra-rich Coors family to tell us what Hispanics in California and Idaho are thinking? Or do they ask Hispanics in California and Idaho what they are thinking? Or what? Do they ask an instant talking head expert in Virginia to tell us what Iraqis think, or do they ask Iraqis? Do the major news networks devote hundreds of hours of news broadcast time to presentation of Republican primaries and convention meetings while providing only a few minutes to discussions in the Green Party about what the major issues are? Or what?

In June 2003 one could quite clearly see the selection process at work in choosing a Presidential candidate for the Democratic Party. As the mass media reported on the debates and public appearances of the field of nine or so presidential hopefuls, the cameras and reporter comments kept fixating on Dean and Kerry, with cameo appearances of Reverend Al Sharpton (no doubt to avoid the overt appearance of racism). Commentators, from their fixation and conversational slant, pretty much have Dean and Kerry "selected" as the only viable "real" candidates, the others all suffering from the tootoo's, either too young, too female, too progressive, or too old fashioned.

In terms of mass national support, almost all entirely generated through the internet, the most well known and active candidate, the progressive-populist Kucinich, was almost entirely ignored by the mass media. It is apparent that the mass media has already decided that populists like Kucinich and Sharpton are not popular in editorial offices and accordingly will never be afforded the status of being a "real" candidate. All this before a single primary vote by real people has been cast.

Those who wish to survive the coming economic crash on their own terms need to "see through" what the media purports to tell us is the public agenda, and who is in and who is out, through their selections. This is of paramount importance because the simple reality is, I am sure you already know, the American Mass Media has a lot of stuffed-shirts with big egos who simply don't know what they are talking about in a great many instances.

During early June 2003, they began in a great herdlike movement to tout the stories of the Bulls, telling us that the economy is recovering. But by the end of June, the Federal Reserve lowered interest rates once again, the 13th time since the collapse of the bubble. Surely the economy was NOT performing well, the Bulls were far too eager to sell to the naïve, and great portions of the more uninformed mass media were clearly too eager to pass on ill-advised "toots".

This points to why the "selections" of the mass media must always be critically examined. The educational attainments of a great many anchor persons, show hosts, and editors is no better than the great mass of people. They can be as easy to manipulate as your next-door neighbor, if they think they are being told something authoritatively by legitimate sources. Should their "selections" have any greater importance than your next-door neighbor's? Why should you be led to a foregone conclusion by your next-door neighbor's news picks?

The truth is that there are a large number of highly educated people with a great understanding of the world and human psychology who COULD do those mass broadcast jobs on a higher plane of accomplishment, usually more astutely than is spewed out in most TV headline news. But such people are not in the media to provide the perspective to properly select and define the headlines with greater fidelity to history and the true significance of the events. Have you ever wondered why? Quite obviously they have been "not selected" in favor of what "somebody" wants to present to you as "what you are interested in" from "people you trust", or at least have come to tolerate.

The more one can see through Mass Media selections, personal agendas, and the limitations of its personnel to discern what is going on, the more one can become free of mass media manipulation. The more one can see through the filters of the reporters (their dreams, schemes, hopes, fears, insecurities), the more one can play the economic game independently and objectively in accordance with what is truly going on.

From this, one can learn to see the "Bull" play for what it is, whether it is in stock market toots or in war drums, and thus one can play for the best of times and the worst of times in accordance with one's own light.

The Slant

One of the most important things to get is "the slant". The "slant" is omnivorous and is ever pervasively seeking to influence you into a certain way of thinking and acting. But, like fish in water, we are often completely oblivious to the currents in the media sea which are pushing us in certain "slanted" directions.

Fortunately, with the advent of Mass Broadcast Media 24/7 on so many channels, the slant is occasionally so over-played that it becomes grotesquely obvious. From the "code words" in the "refrains" of the overplay we can sometimes see that the "news" is often less straightforward reporting on what is happening than it is the telling of grossly biased and slanted "stories" which sell a "Line". The "selling" of the invasion of Iraq to find the weapons of mass destruction comes most easily to mind as a perfect example of such a "slant".

The slant has always been with us but it is now far more dynamically important in today's world. With "global news" in a virtually instantaneous electronic village, the play of the news on important issues sometimes has far more impact than it had in the previous "print" generations. The "electronic slant" can whipsaw up a panic in very little time, even among billions of people virtually overnight, as was recently seen in the SARS episode. Accordingly, we must deal with the "slant" in more dynamic ways to counter-act its negative effects and its tendency to hyperinflate hysteria.

The SARS slant: an epidemic spreading, here, there, out of control, killing 4% or 15% of the those who caught the virus, damaging the economy of Toronto and Hong Kong for a few weeks. The reality: medical scientists still weren't certain about whether most of those cases were caused by the same bug long before the press lost interest, and, regardless, the true risk was never fully understood. Some health care specialists reported that successful, inexpensive treatments were highly effective but the media was not around to broadcast this information because the media was not technically competent to analyze competing claims.

Based on 'the slant", you could easily have been led to take a Bear response to SARS by assuming that the economy was going to go to hell. But you probably would have wasted your time. The moral: get past the headlines, use the internet-media to get other points of view, and realize that the Mass Broadcast Media is called the Mass Hysteria Broadcast Media in this special report to make a profoundly important point. The Mass Hysteria Broadcast Media is as much a "shock em and rock em" entertainment venue as it is a source of VALID information.

The medium thrives on high impact threats and events about dangers which rivet people to catch the latest soap opera dose of "As The World Turns". Or Wolf Blitzer's, TA DA, "The War In Iraq" (with musical crescendo), SARS, Anthrax, Tarot Snipers, and "The Empire Strikes Back" at Iraq are perfect suspense thrillers which generate real emotions and even panics. These stories are perfect for generating huge repeat audiences for the mass media and that is why you hear about such things incessantly, whether the minute by minute reportage is "news" or no.

Is the Mass Broadcast Media biased to over-generalize, over-report, and over-sensationalize? Absolutely. It means bucks, big bucks from advertiser revenues for the huge repeat audiances the "panic slants" generate. The Sorcerers now even use musical crescendo themes to induce a greater emotional impact.

It is easy to see that this makes our lives highly prone to panics and hysteria. The omnipresent repeated "slant" in the reporting on SARS served more to feed a panic than provide real information to the public about what was really going on with various scientific responses to SARS. Consequently, people grossly over-reacted, not on the basis of the facts, but on the fear which was being transmitted through the media.

This example is easy to see, but many Americans have difficulty seeing that the same thing is true for how they were stampeded into supporting the invasion of Iraq. The plain truth is, as most of the world knows, that the facts dished up about Saddam's threats, weapons, and links were "vaporware". The fear-generating "slant" communicated through the Mass Media about Iraq and Saddam spooked many Americans into stampeding into Baghdad. With a schmoozing, know-it-all attitude, the Imperial Faction turned the Mass Hysterical Broadcast Media into willing accomplices for transmitting enough "emotive" push to march a huge nation into Iraq. There wasn't a single fact, absolutely no real threats from anyone, that motivated the invasion. Not a single one.

Omissions and Spells

To take it again from the top, the merchants of the media, especially the Mass Broadcast Media, continually define what IS the news, what is NOT the news, what "people are interested in" and what is the "legitimate take" or "slant" of the news.

It is quite clear that we are being manipulated all the time by an editorial class to be aware of certain things and to be oblivious of other things. One of the most powerfully important techniques they use is "soaping" us with endless replays and updates which sustain the "suspense thriller" spell and generate a morbid sense of connection with fear, panic, or outrage over events.

It is doubtless true, as many professional journalists claim, that the average reporter usually works hard to provide an objective report. The problem, as professional journalists have to admit, is that a lot of the qualifying context and perspective often gets left on the cutting room floor. Some important and wonderful stories ready to air are never played or are never played in appropriate fullness. This brings us to the "omission", which is one of the most powerful editorial forms of manipulation.

Virtually omitted from North American broadcast news, for instance, are stories about events and conditions in Africa, India, and Latin America. Okay, maybe every few months something on AIDS and maybe riots or floods in Latin America Do you wonder why? Well, if you are not descended from people who came from those places, look in the mirror and the answer will be staring you in the face. The Mass Media thinks you are like them, that you are not interested in these inferior places. Britney shaking bootie is far more interesting.

To rationalize its sins of omissions, the press buffaloed people for a long time with excuses about the demands of limited time and space. But this is now just an insulting put-down. As time has gone by, it is easier and easier to see that more and more of this suppression serves conscious agendas of manipulation more than it serves the limitations of the press in reporting "objective facts".

This is easier to see now that there is so much time to fill. We clearly can see that the various news channels sometimes have great difficulty in filling space with "THEIR" agendas. We can see they end up stuffing puffy nonsense about wrinkle cures into nearly empty spots. Or they spend way too much time with phony, staged events which are not news at all, such as another five minute recital of threadbare "soundbites" by old "Wartime" Bush. And of course, you can also count on catching a snatch of a Britney-look-alike shaking bootie for those off-moments.

During 2003 and 2004 the most omitted story line in America will be the rise of the populist progressive agenda. The mass media, especially the embeded lap dogs who "think" they need to be able to talk with office holders to write a decent story, will persist, in the face of all facts, in claiming that American politics continues as normal and that the populist agenda, which most likely is widely supported by at least 30% of the population, is too radically fringe to even deserve press coverage.

"Presto", the Sorcerer's of Broadcast wave their remote control wands, "they are of no account and do not exist!"

Fortunately, especially since the advent of the late 1990's, great numbers of people are now waking up to the fact that many of the primary "news" reports of our times are less real "realities" than they are carefully stage-crafted "acts" which are created through the manipulation of appearances and utterances. These "acts" are quite often stripped of context, repackaged for soap appeal, and are broadcast and repeated endlessly through the global media, serving to cast "soundbites" and "images" into... something akin to hypnotic spells.

In other words, our world and our history is being shaped less and less by " real events" and stories told about them in the plain ways of old fashioned reporters, intellectuals, writers, and editors of the print media. More and more our world is being shaped by instigators and producers who serve up "spells" with hyperventilated multi-media effects, rather like sorcerers seeking to conjure reactions.

God help us all.

The Medium Is The Message

Nearly all of the print media everywhere, and not a little of the broadcast media, were originally organized to enable the publisher and the editors to speak to the public so that they could influence the conduct of civic affairs. In this sense, it is all slanted, it is all an attempt to influence you. That is the objective truth. It is that simple.

The basic reasons for all of this are pretty simple. If you want to control the outcomes of politics, the destiny of the nations, and the regular course of economic development and prosperity, you control who speaks. That determines who is perceived to have social standing, power, and legitimate opinions. This automatically sets what the agenda of discussion will be.

Hence, since the earliest times, the real struggle over politics is over the right to speak and, equally important, the right to hear. There is no other real mystery to either politics or economics. If you can control or influence who can speak, where, when, and how, you also influence who can have access and hear about what is going on.

This is so simple, nearly everyone understands these things intuitively, which is pretty much the level at which politics is organized in America. Most forms of social organization and governance are carefully organized to provide rules of inclusion (who can speak) and exclusion (those who are not permitted to speak), and these rules are usually strongly influenced by the methods of communication, the medium if you will. These communication principles are pretty much the social equivalent of Einstein's "Theory of Relativity".

From the history of the past several hundred years we have learned a very interesting corollary to these principles. If you change or shift the means of communication, such as the technology, you change the scope of who is included or excluded and hence you influence the outcomes. If you can influence the outcomes, you can lead people to a particular future.

It begins with Face-To-Face communication. Face-To-face communication is highly effective and almost all personal decisions which matter are made this way. If you control or influence who can show up to a meeting, you influence (you can bias) what the meeting will talk about or how the topic will be discussed. This of course affects the final decisions.

For an expanded audience, print media is essential for disseminating information, but it is linear (one way), very slow, and is not directly interactive nor helpful in actually making most decisions for any group. Hence decision-making by this method is limited primarily to the occasional casting of ballots.

Radio and TV are far more flexible and thus interesting, they can provide both linear broadcast and interactive discussion. They can do both at the same time, they can broadcast face to face decision-making of a small group to a large group of people. This is in essence the kind of media-induced social structure we have today.

Control or influence any aspect of any of these methods, you influence the outcomes. Indeed some of the most important economic and political struggles of the past revolved around a struggle to create or control the methods of communication. Guaranteeing the "Freedom of speech" was essentially the outcome of a struggle about breaking state monopoly of control over who gets to communicate what kinds of things. 'Freedom of Religion" is a variation of this struggle.

Through the ages this struggle has taken many forms and indeed the struggle is mutating again right before our eyes during this strange beginning of Century 21. The transmutation is occurring principally on the internet, which melts down the communication of all information into nearly perfect relativity where space, time, and money mean very little. This melt is changing EVERYTHING, and these changes will accelerate during the next 15 years more rapidly than the changes induced by TV during the past 30 years.

The awakening to this kind of awareness about the power and the influence of the media began on a mass scale with Marshall McLuhan. He coined the term "the medium is the massage" during the Vietnam War when Mass Broadcast TV Media came into its own. This was when "the six o'clock news" on NBC, ABC, or CBS became the principle means of serving up the headlines which groups of journalists thought were the most important stories or datums of the day. Prior to that it had been radio's role to serve as a sort of news ticker tape for the prior generation. But even during the heyday of radio, the daily metropolitan newspaper provided the real facts and background perspectives to the people who wanted the "real news".

With his adage about '*the medium is the massage*", Marshall McLuhan was trying to teach people to observe how the new national medium of TV had already begun to fundamentally change American culture, was shifting its agendas, who was talked about, who was seen, who could be successful on TV and how, as well as change how politicians went about creating events and talking to people.

TV was fundamentally different, he argued in his books, creating completely different political results. And most importantly of all, he observed, it created for the first time a truly NATIONAL CULTURE, and probably eventually would create a GLOBAL CULTURE. Regional cultures flourished when large metropolitan newspapers were the primary MEDIA for news. Now a national culture was emerging through the electronic wizardry of the TV networks. Inevitably, a global culture would arise in the future.

During the 1970's, when McLuhan's work became known, the future for this national culture looked very positive. But this was before the proliferation of dozens of channels of junk TV and emergence of the Hollyweird Sleeze Industry, which primarily show how ugly culture can become by broadcasting content which celebrates extreme violence, seriously-challenged ethics, or how boneheads can become rich and famous by shaking their bootie.

So during these first few years of Century 21, we are not nearly so inclined to innocently suppose that the Mass Broadcast Media is a benign one. At minimum, we know we need to begin with a highly skeptical approach. As a necessary complement, we cannot avoid taking very deep draughts of pessimism about the infinite corruptibility of power.

Good Citizens, Mass Media, Jingoism, Manifest Destiny, & Madness

To truly understand the powers and the shifts in the global media, let us begin with how the media makes fools of Good Loyal Patriotic Citizens of the type who complain loudly about flag desecration and free speech for celebrities like the Dixie Chicks, heavens forbid, or even French Presidents, who happen to disagree with what the big shakers and makers are pushing as the daily "Big Line".

This is not difficult, even academics and journalists will understand it. To view the power of media sorcery over our "Good Citizens", we need to look backward into the destiny of Manifest Destiny. From this we will see how history has been seriously manipulated by those who command mass media.

Our patience in plodding through a brief tour of this history will be greatly rewarded with liberation. The tour will give us the insight we need to plainly see the low level of crass emotional sorcery which is used perennially to steer the ship of state. This will both free us and alert us to how the corporatized Mass Hysteria Broadcast Media may use its powers during the next several years to stampede people into the next daily "Big Line".

We begin with the essence of history. People. Everywhere, most people genuinely are the same in what they want. Doesn't matter what culture, race, continent, not even income level. People want strongly to be "good citizens" or good members of their family, their community, and their tribe. They want peace, freedom, prosperity, and respect and they are willing to give a lot of effort to others to help make these things possible. Accordingly, they will, if directly approached by their leaders, bend over backwards to cooperate. This being so basic and so simple, it is truly hard to understand how it goes so wrong.

Let us begin with one of history's clearest examples about how it goes so wrong. During the 1930's and even well into the 1940', the mass media grabbed "Good Germans" and led them through history's worst meatginder. These "Good Germans", the kind of industrious, democratic, sincere, well-meaning people who are wonderful members of any community, were manipulated and led, via mass media technology, to support one of the most lunatic crusades of all time.

It is true, the Germans at the time had their little "weakies", including Martin Luther's rabid anti-Semitism which the Lutheran Church had browbeaten into Germans for 500 years. But that is not why the crusade began. The crusade was not based on the weakness of Germans but on their best points and values.

This is a really important point which you must really get. Almost all the academic historians out there, at least those who have written in English, write stories and books which analyze any number of types of "weakness" and evil dark side in the Germans which allowed the Nazis to gain power to wreak havoc. They have it all wrong. The Nazis rose to power by learning how to manipulate the good side of the Germans. That is what makes it all so truly tragic.

The catalyst was a strange band of neurotic personalities who were obsessed with power and domination, led by a man who had a strong, confident oratorical style. They seized upon the latest mass communications technology of radio, brand spanking new, to convince "Good Germans" that Bolshevik communist Jews from Russia were engaged in a campaign of sabotage to disrupt their government and enslave them in poverty under a despotic international world order ruled from afar.

Do you recognize this conspiracy theory after you remove the German content? Yep, this is whence it came. The American right wing later copied it nearly intact to sell the Cold War against, who? Bolshevik communist Jews from Russia. Promising on a daily basis over the radio news channels to lead Germans into a new 1000 year era of security and world peace which would finally make the world safe to fulfill the "Utopian Dream" of "The Volk" (the people) for justice, equality, security, and prosperity for all, Adolf Hitler and a small group of Nazi strategists began to promulgate the "New German Millennium". Obviously they were thinking in much larger terms than a "*Project For A New German Century*". The "New German Millennium" was to be nothing less than the final heir to the ancient Roman mantle of World Empire. In the One World German Superpower, history would end in peace and shine everlasting glory on the Volk.

Is this beginning to sound suspiciously, uncomfortably familiar? It should and if it doesn't, you must read again about "*The Rise of the Imperial Faction*" in Section 3. If you think it's uncanny, I think you are on the right wave length.

Better yet, read the "*The Rise and Fall of the Third Reich*" by William Shirer (originally by Simon and Schuster, 1960). It's all been done before, sorry PNAC, the history copyright people are coming to revoke your corporate charter unless you give credit where credit is due to the Reich Ubermensch (Nazi Overlords).

With their new powers of radio mass persuasion, which we might think of as variants of hypnosis and emotional sorcery, some of the most brilliant but twisted minds of all history demanded more and more power from the people and their Constitutional Republic. The Nazi Reich Ubermensch progressively consolidated all authority into a police state which efficiently and directly led the "Good Germans" with great haste into one of the deepest tragedies, if not THE deepest tragedy, in all of human history.

In many ways it was the same with Japan and Italy, just a little less dramatic. To see how the same "story" was played to these people, we need to become pre-war (World War II) Italians and Japanese for a moment and see the world through their eyes.

Many Japanese were convinced that "white racism" had at all costs to be defeated militarily to safeguard the oriental peoples from imperial enslavement. Accordingly, they used the same mass powers of communication and hypnotic sorcery to mobilize earnest, sincere, hard-working people to undertake a vast program of military expansion. Ironically, much like the Germans, these hardworking Japanese thought the entire time that they were creating a western style democratic state under their Parliament and Constitutional Monarch (titular). They thought they were struggling during a time of great sacrifice to safeguard themselves from the Imperial New World Order being imposed worldwide by white racists from Europe and America.

For the Italians, it was a similar sort of nationalistic, patriotic fervor which drove them to attempt to rebuild themselves proudly as a new nation with a real parliament which unified the entire country under Mussolini and a new Constitutional Parliament (titular Monarchy).

World War I had nearly destroyed Italy's late 19th century re-unification. Prior to that, Italy had not been a unified "nation" for many centuries. Italians were eager to hold onto their re-unification, which had brought most of the social and economic progess during the past 50 years. They went with Mussolini to hold unto their unity and the progress it had brought. Mussolini easily convinced them that driving the imperial forces of Great Britain out of the Mediterranean Sea, which were constantly meddling in the economic affairs of all the nations of the area and imposing one-sided treaties, was long over due. They were proud to join a coalition of nations to get rid of the robber barons who threatened their prosperity and well being.

Looking at history in this way, from the point of view of the losers, we can see how it is that perfectly Good Citizens are all too often led into horrible misdeeds and terrible tragedies which linger for centuries in the history books and in our souls. We can look into the well-meaning aspects of the intentions of these 1930's people in a sympathetic light without exonerating the terrible wars and crimes committed by their nations and leaders. In the same way we can also understand how "Good Americans" stupidly but idealistically committed such horrible crimes in Vietnam.

Seeing the world in this way gives us historical insight, a long vision of causes, reactions, collisions, and eventual effects. We need to see the world correctly in this long view so that we may avoid perpetuating the stupid and costly reactions of the past.

Even as with these earlier eras of time, George Bush Junior, the Imperial Faction, and the Mass Broadcast Media in the U.S. have once again proven the simple elegant truth of Hermann Goering's proposition. Once again we have been shown how to use the powers of mass media technology to turn fear and the good intentions of well meaning "Good Citizens", into a process of consolidation of power and military adventurism IN THE NAME OF protecting and enhancing the great "Utopian Dream" of the "Volk", er, the "Dream" of the American People.

It was all too easy. The American character, like most others, has always been easily led to channel its anxieties onto foreign adversaries. Foreign wars would appear to be difficult for Americans to undertake.

At first blush, it would seem that it is very difficult to get Americans to undertake foreign military activity. The U.S. nominally has a Constitution which virtually strangles public decision-making. It is purposefully NOT democratic so to prevent military adventurism by demagogic politicians and special cliques. Its original leaders gave the strongest of dire warnings to avoid "foreign entanglements'.

Yet despite all this the American character has been led repeatedly into aggressive campaigns against other peoples. And almost every one of these campaigns have been inspired and cheerled by the media of the day. The inspiration has almost never been generated by a real attack on the United States (World War II being the notable exception). By raising up a great furor about grave dangers to American liberty, rights, and territory, the media easily created a consolidation of opinion which coopted the man on the street sufficiently to overcome the crabby requirements of the U.S. Constitution.

This Constitution requires such a tendentious due process for making war that it is difficult to get people to agree to make war except under the terrible emergency of an actual invasion. When self-defense from real violence is necessary, people usually know what to do, the law is merely formality to which proper lip service is easily given. And indeed, the Constitution gives the U.S. President very little limitation in acting to repel actual invasions or in defending the preeminence of the Constitution as the highest law of the land. A war of self-defense is understood to be a "special case" which requires the President to act with great dispatch and with full authority in the manner he/she thinks best.

And so it was that the American Empire was created by Presidents purportedly acting in self-defense. Generation after generation, the media sold "The Line" about threats from abroad and an invasion of American rights, interests, and/or territory. The President dutifully marched his troops.

Manifest Destiny I

It began with the campaigns to extend the power and territories of slave owners in the Americas, all the better for a nouveau commercial class to establish for itself the "Utopian Dream" which only Barons, Dukes, and Kings had enjoyed in previous centuries. First it was a matter of clearing out the so-called marauding and lawless Indians in the Southern States, the better to make civilization and economic development for Slave Owners. And so Andrew Jackson cleared out the Indians from their ancestral homes and sent them packing to Oklahoma. For good measure, his minions practiced new arts of bio-warfare by giving the Indians blankets infected with small pox.

Think about it. Andrew Jackson became a President after the first use of bio-weapons by Americans to decimate people who were in the way of the expansion of the "American Dream" of that time. The face of that racist still sits on the U.S. $20 bill. Think about that every time you pull one of those out of your wallet. Andrew Jackson, part of the "Evil Axis" of terrorists who use weapons of mass destruction.

Then it became a religious struggle in a war of words which ended up dividing the country into a bitter civil war . Regional papers argued their positions on who should be entitled to grow the United States across the continent, the slavers or the peasant farmers from Northern Europe. We ALL know where that ended. The peasant farmers got to kill more Indians in the West, but, thanks to the religious media, at least slavery was illegal. The crimes of the 19th century continue to shape the world even to this very day. In the end, everyone lost in that terrible tragedy. Even the children of the peasant farmers.

These were the precursors of Manifest Destiny. After that, it was go for broke with Manifest Destiny I, America The Great, from sea to shining sea across the entire continent. The "only good injun" along the way was a dead one. Again, the regional and local newspapers sang the sagas of the conquerors. But this time new voices and media added great power to the push. The publishing industry expanded into the national distribution of books and magazines, which aggressively sold "The Line" about the wonders of the West, the greatness of the pioneer movement, and how barbarian, stupid, and unworthy the Indians were. A good portion of that media is still around.

To this day, the Americans cannot quite get it okay in their soul to do justice to those their forefathers sought to exterminate. It is truly amazing how few want to look back and fess up. Americans perpetuate over one hundred years of inertia in looking the other way. The crystallization of negative emotional attitudes which defame and divide people, all fueled and fanned into raging fires by publicists of the new pulp media of the late 19th century, can still be seen and felt. In many areas, Americans are still allowing the mismanagement of Indian treaties and trusts in favor of greed and narrow self-interests.

Manifest Destiny II

By the end of the 19th century, Manifest Destiny I was passe. The nation was covered from sea to shining sea with daily newspapers linked together with telegraph lines. Within 20 years or so of this nationwide linkage, Randolph Hearst and other 19th century jingoes invented the American war against Spain to extend the reach of Manifest Destiny.

The last few years of the 19th century was the perfect time for Hearst and others to extend the boundaries past North America where all the property lines had been staked. The pioneer settlement on the continent was done, Canada was out of bounds because the British Empire protected it, and a land war to take over Mexico would doubtless be extremely costly in lives, and, well, rather imperious.

But the tottering old Empire of Spain was easy pickings. No one cared about its imperial claims. A despotic dynasty had impoverished both Spain and its incredibly seedy colonies. The Spanish had lagged so far behind in industrial development that Spain's ability to actually wage war was in serious doubt. Accordingly, now was the perfect time to grab its colonies and liberate them before other imperial powers attempted to heist them (which was probably inevitable).

Liberated from Spain, Cuba and the Philippines could be remade in a manner more suitable for the American values of grabbing whatever was around from the natives. In little time, free market capitalists could build a civilized world of corporatized hereditary fortunes to replace the hereditary aristocracy of the old Spanish empire.

Best yet, by overturning an evil tyranny, the promoters of Manifest Destiny could sanctimoniously claim that they were not imperialists, they were just "Good Citizens" serving as liberators. This "slant' worked quite well then to pull the wool over the eyes of "Good Americans". Indeed, much territory had been forcibly "liberated" from Mexican tyrants during the preceding 50 years by rallying "good citizens" to the side of Manifest Destiny. It had always worked and indeed it still works amazingly well, as demonstrated most recently with Iraq.

So it was that Manifest Destiny II began to play on the telegraph wires which linked the first mass media consciousness of the American nation. You were no longer from Vermont or Kansas. You were now an American, and Americans were now acting on the world stage while ripping off, er, liberating some provinces from a tyrannical, decrepit old empire.

It was easy. A U.S. Battleship, the U.S.S. Maine, mysteriously exploded near Havana Harbor. Soon Randolph Hearst had headlines playing day by day about evil Spaniards, the despotic rule in Cuba, the threat the Spanish Navy poised to American free trade in the Caribbean, the heroic efforts of American Theodore Roosevelt and his "volunteers" to help the Cuban people free themselves from their cruel and tyrannical overlords, etc. This story "Line", and others like it, played well, Hearst vastly expanded his circulation and his chain of newspapers. The U.S. government picked up control over a large number of islands.

Manifest Destiny II continued to play out on the telegraph wires quite well through the 1930's and then it flourished equally well on radio. As American gunboats patrolled Central America and the Far East to make the world safe for American plantation owners, Americans supposed they were occasionally preserving the rule of law and helping protect impoverished peasants from bandits and violent rogues. Over and over again, American troops and naval vessels intervened in various Central American nations to protect so-called American interests, and Americans scarcely noticed that they had been unofficially dragooned to support yet another small war. They were unable to see that the biggest bandits marauding Central America called themselves Anglo names like "United Fruit" and lived in New York and other cities in the U.S. Or maybe they saw it and became totally marginalized into the same fate of "irrelevancy" which has befallen all populists groups in the U.S. who opposed large concentrations of power and foreign wars.

World War II more or less ended Manifest Destiny II. A fortunate partnership between two of the most influentual leaders of the previous millennium, Winston Churchill, conservative Prime Minister of Great Britain, and liberal Franklin Roosevelt, President of the U.S., brought together an interesting union of the highest principles of conservatism and liberalism to remake the politics of the world. Under their tutelage, imperialism was made obsolete, nations were to be left to self-determination by their own peoples, and interference in other countries was to be taken only to prevent outright anarchy from gaining the upper hand. Racism, Manifest Destiny, Imperialism…all dead, gone, done. The United Nations now would preserve a world community in equality and peace as best it could.

Despite these best of intentions, the stand-off between the Soviet Union and the United States fractured the United Nations and largely prevented it from functioning as an agent of preserving the peace during much of its history. Soon, machinations of the communists were met with machinations from the Western powers, primarily the U.S. and Britain. For the most part, the two armed camps contrived to paralyze each other from overt foreign aggression by balancing nuclear terror and "Mutually Assured Destruction".

During this long M.A.D. stand-off of some 40 years, the ambitions of Americans turned inward into the creation of continental corporate empires and the rising prosperity of a vast supporting middle class to manage them. Acquiring new opportunities through land and resources no longer mattered. It was industrial technology and infrastructure development which mattered.

The Americans succeeded quite well in their industrial development using the "New Deal" idealism and principles which the Democrats installed under Roosevelt and managed under Truman,

Eisenhower, and Johnson. Roosevelt and his Democratic alliance more or less locked the economy into prosperous support of middle class interests and needs. Eisenhower, though a Republican, had no interest in challenging the success of the coalition and in fact supported it handily by laying the foundations for the vast interstate highway system and by seeking strongly to find peaceful ways of dealing with the nuclear nightmare under a (desired) collegial relationship with the communists.

Under the New Deal, corporate interests were NOT to be preeminent and indeed they were not. Their power and behavior was to be carefully regulated to insure good citizenship. Under the sharp-eyed regulation and guidance of national agencies, corporate activity actually worked to uplift the nation for some 30 years until the advent of Republican presidents, beginning with Nixon.

The New Deal alliance was the first tacit attempt in the United States to institutionalize economic cooperation between all members of society within American government. It was, so far as it went, and despite its shortcomings, rather successful. The New Deal Alliance won World War II over all other imperial powers, brought the U.S. to its diplomatic preeminence in the world by insisting on the values of equality, freedom, and cooperation, and it gave rise to nearly 40 years of unparalleled economic success even as it stood off a massive international mobilization of military power under the Bolsheviks.

The success of this great Democratic coalition included prosperity for a rapidly growing middle class, social security, the generation of mountains of hyperbole about the American "Utopian Dream" for all people, and the successful implementation of far reaching social change, not the least of which was the end of most overt racism.

All this was ministered by a watchful print media which undertook the role of "referee" between the claims of government, corporations, political parties, unions, and various groups of citizens. For the most part, the press supported the premise of the Roosevelt coalition. Namely, that all of the basic economic interests, including the citizen, deserved a place at the political table. Accordingly, the print press provided room for a diversity of viewpoints across the economic and social spectrum. To a great extent, the print press still does. It is another matter altogether for the Mass Broadcast Media, which is progressively narrowing viewpoints and values to a small boomer go-go bubble dominated primarily by East Coast attitude.

Manifest Destiny III

Most tragically for Americans, the great success of the New Deal coalition and their triumph over World War II made them dangerously susceptible to the play of new forms of Manifest Destiny. It came much too quickly in the form of a vastly oversold "destiny" to stand off communist take-overs of foreign countries.

It began of course with the Korean War, which was undertaken and fought in accordance with international law and the ideals of the nascent United Nations. The Russians boycotted United Nations discussions of North Korea's invasion of the South and this allowed the United States to weld together a United Nations "police action" to defend South Korea. This worked out politically but it was immensely costly in lives. The armistice from that police action still stands and South Korea has become one of the "wunderkind" economies of the world.

Despite the cost in lives, it seemed that the idealism of Churchill and Roosevelt had succeeded in saving South Korea from communist conquest. At least for awhile, it seemed that Manifest Destiny had been converted into an international citizen. Americans had become global citizens working within international law to maintain law and legitimate governments. But unfortunately, the Russians never repeated that "boycott" mistake and the United Nations was never again able to defend against an invasion until the Soviet Union was dismantled.

After the Korean War, Mass Broadcast TV Media began to supplant both radio and the print media during the 1960's and 1970's. As these new eyes on the world began to develop, presidents Eisenhower and Kennedy ordered a few small defensive police actions to preserve and protect weak or unstable governments from outside encroachments. These were all understaken outside of the United Nations because of stalemates in the United Nations.

Since the United Nations was stalemated by the Soviet Union, the protection of governments and nations against invasions by communist forces could only be undertaken by individual nations. But this was as difficult to obtain as a U.N. Resolution. Korea had turned into a quagmire which cost some 50,000 American lives. The World War II generation in control of American government was weary of war and wanted no more such direct interventions.

And so the "proxy war" was invented. Instead of American forces being used directly, which would involve a lot of political controversy, the U.S. government would provide "aid" and "advisors" to foreign governments to fight against communist insurrections or invasions.

That was the theory but in practice it was usually perverted into aggressive invasions to overturn legitimate revolutions which had ended corrupt, brutal elites. As they say, the old ways die hard. During the 1960's, American politicians allowed themselves to be talked into fighting two such invasive 'proxy" wars, in both Cuba and Vietnam.

In Cuba, Castro threw out a brutal, crooked Mafioso family and after about two years in power turned to the Soviet Union for support and defense against hostile North American corporate elites. In Vietnam, Ho Chi Minh successfully forced the French to withdraw their plantation lords. He was in process of creating a national government for Vietnam but French and American interlopers refused to honor the electoral process which had been set up. The negation of free elections in Viet Nam met immediately with armed resistance from Ho Chi Minh and his communist supporters.

From these struggles, two terrible tragedies were born. A small aggressive U.S. backed effort by "proxies" to invade Cuba and overthrow Castro failed miserably and quickly. Miffed, the U.S. government spent the next 40 years pretending Cuba was not there, hoping that someone would eventually kill Castro.

A much larger effort against the communists in Vietnam also failed but only after terrible crimes had been committed and losses could be counted in every town in America. Both actions were aggressive invasions of "liberation" created and condoned by "Good Citizens" which were highly unwelcome by the overwhelming majority of the people being invaded.

The terrible tragedy of Vietnam for the most part ended visible American involvement in "proxy war". The "proxy" was so corrupt and incompetent, a half million Americans had to do the fighting. Americans eventually realized that they had been duped into the Vietnam War by Lyndon Johnson's Big Lie. He claimed, oh so like the "Maine", that North Vietnam had attacked American destroyers in the "Gulf of Tonkin" with two-bit half-arsed gunboats.

Of course he had to respond... Americans finally woke up to the huge piles of carnage which had been created in support of the illusionary "proxy", the fake little junta of South Vietnam. American's began to realize that America, like Japan many decades before it, simply could not succeed in imposing its will on Asia. Like the Japanese, they had to accept great military loss as the price they paid for ending the conflict. Americans also had to pay for the their jingoism with national humiliation at actually losing a war.

About the same time, racism against blacks publicly died. After that, American jingoism played very poorly in the mass media. Imperialism seemed as dead as the old racism and the overt play of Manifest Destiny. Even Khomeni's humiliation of the U.S. embassy personnel during the Iranian Revolution failed to produce aggressive military activity, though the American right wing still defames Carter for "allowing" the Iranians to get away with it.

A succession of Republican presidents from Nixon onward cobbled together a coalition of conservatives, ultra-right ideologues, religious authoritarian fundamentalists, corporate interests, and nascent but uneducated additions to the middle class. Beginning most auspiciously under Nixon, the apostles of greed began to achieve ascendancy under the Republican presidents. They progressively corrupted the political and economic system to serve narrow elites by destroying more and more of the regulatory balances which had been put in place under the New Deal Alliance. Nixon of course was later eventually forced to resign for dishonesty, as indeed many apostles of greed were forced to do after the breaking of the dot-com bubble.

As the U.S. government began a massive shift to serve elitist interests, the good-old-boy Manifest Destiny crowd kept itself alive deeply entrenched in the CIA and State Department. It went underground and festered in a succession of limited covert support of "proxy" wars, mainly under the Republican presidents.

Under an aggressive, often hostile press, which was still dominated by old-fashioned print, the "slant" to intervene directly in other people's civil wars was met with a great chorus of skeptical hostility. Even Ronald Reagan, one of the most personally popular politicians of all time, had to walk a very careful line to avoid overt personal involvement in foreign adventurism with direct intervention or "proxies". Scandals over American secret arms deals with Iran, and covert drug running to finance counter-revolutions in Nicaragua and El Salvador nearly engulfed him.

The jingoists could not directly incite Americans to intervene militarily in Nicaragua and El Salvador. The best they could do is use gunboats, planes, and Marines already stationed in Panama to "arrest" the ex-CIA double-dealing drug tyrant of Panama. Along the way they also forced out the hated Duvalier stooges from Haiti, which the U.S. government only most reluctantly undertook. They also ousted some Castro stooges from the small Caribbean island nation of Grenada who had actually carried out a take-over "putsch" with only a few hundred men.

The old jingoism seemed to be slowly dying out along with racism. It seemed for some fifteen years that only a few dying embers were still smoldering in the dark recesses of the CIA and the basements of the Pentagon.

All the other covert CIA "activities" were rationalized as defensive actions to support foreign movements and governments against communist plots. The infamous Dr. Kissinger, who is sure to go down in history as one of America's more bizarre characters, was a master of rationalizing these connections. His involvement in masterminding covert dirty tricks in a number of countries, which resulted in right wing death squads murdering thousands of people under Fascist governments, has earned him long-standing indictments from democratic governments which would dearly love to throw him in jail.

The fact that Mass Broadcast Media in the United States can continue to provide this international criminal and consciously dedicated, paid foreign agent with a public platform on TV is excellent proof of how historically naïve and uneducated the boomer editor and news anchor class really is.

With the exception of Iran and Nicaragua, most of the covert intrigues played with very little interest in the press. When the press did cover aspects of these stories, it was with very little fervor of support and with large measures of questioning and criticism of covert American engagements and the failures of American policies, both past and present.

The "good citizens" of Manifest Destiny III even learned to begin to atone for racism and some of the errors which jingoism had created in the past. They matured to live with the fact that other peoples would occasionally reject, even with great force, American and European world views and demands.

During the Reagan years, Manifest Destiny was publicly simplified into standing tall with Reagan as a new day arrived while America rebuilt U.S. military technology and outspent the Soviet Union on REAL weapons of mass destruction. With Star Wars defense programs, Manifest Destiny seemed destined to go into orbit rather than fight more wars on Earth. But even this "last form" seemed to fade away. After Reagan made a deal with Gorbachev to end the nuclear arms race, Manifest Destiny seemed to be on the way out. With the subsequent dissolution of the Soviet Union, it seemed all over. The final forms of international competition and struggle over world domination seemed to be dissolving away , dissolving into amicable agreements.

Bush Senior was the perfect Republican successor to Reagan's ending notes. With great understanding of the world, garnered from a professional career with the CIA, he straightforwardly conducted the Gulf War to protect Kuwait from foreign aggression. Seemingly, he kept in full accordance with international and constitutional law. He was the first American President since the Korean War to make the United Nations work as the ultimate international "cop" to stop an invader.

Because it was a "just war" without the baggage of the past, the Mass Broadcast Media, as did all of the media, made much of the War. They found it popular to popularize it. Through CNN's coverage, the 24/7 news channel format came to the fore worldwide and showed the power of global linkage.

On the open battlefield, the U.N. program worked. It seemed that Manifest Destiny was truly dead and that jingoism had little room left. But U.S. covert manipulations failed. Undercover operations, managed by the same old hidden Manifest Destiny crowds whose policies and manipulations had wreaked such havoc and massive deaths in a multitude of failed programs ranging from Vietnam, Argentina, Chile, Nicaragua, El Salvedor, the Congo, Angola, Afghanistan, and other venues, promised the Crescent Moon of support to the Shiites of Southern Iraq if they would rise up and throw Saddam out of power. The Shiites rose up, the covert operations crowd welched on the deal, and some hundred thousand Iraqis lost their lives at the hands of Saddam's army. The uprising failed miserably, leaving recriminations and karma which are still coming home to roost, like so many other failures of Manifest Destiny's covert operations department.

These recriminations were covertly swept under the rug, but now they're back! In Iraq today Shiite leadership has returned from abroad and is vociferously lifting up the rug and telling everyone just what they think of America's Manifest Destiny Meisters of covert interference. It is not pretty to hear and it may yet cost America dearly.

But we are getting ahead of the story. Clinton was elected in 1992 after considerable concern began to arise about many sour economic notes which had emasculated the "Reagan Revolution" of self-reliance and de-regulation of the economy. With his confident talk of supporting the vigorous expansion of the new technology economy within a new era of international cooperation, Clinton seemed like the ideal president to pull the country further away from the dying embers of Manifest Destiny.

As if on cue, electronic media began to proliferate after 1992 in ways which seemed to break up monopolies everywhere. Technology seemed to bloom into just about everything. The internet began to connect everybody, and the Mass Media began an eight year toot of the seemingly boundless running of the Bulls. The Republicans preached that unlimited entrepreneurial opportunities seemed to be opening up in the new high technology economy, all one had to do is junk government regulation and taxes everywhere.

The Democrats did not do much to oppose the argument. Their main strategy was to maintain a centrist position on retaining enough taxes to pay down Reagan's massive "Keynesian deficit spending program" for military hardware. They managed to keep social security barely adequate but lost the battles on reasonable health care. They caved in on most issues of regulation and social welfare, allowing the old New Deal structures to fall into further decay.

The Mass Media obligingly promoted the latest edition of the "American Dream". As they sang the praises of the unparalleled bull run in the Clinton Feel-Good Bubble, a vast new digital technology began to create an entirely new economics. The media chronicled in great detail the fortunes which rose and they celebrated the apparent liquidation of old concentrations of power in the old industries. Within less than eight years thousands of new companies were created to turn the impressive abilities of the internet into the Bubble of instant billionaire Dot-Coms. The American version of the perennial "Utopian Dream" seemed at hand for ever larger numbers of people.

What could go wrong? Sure, there were environmental issues, AIDS, and violence in Palestine, but all the issues seemed ultimately resolvable. There was nothing truly of crisis proportions. The main sour note was crisis in Europe over the long break up of Yugoslavia. That eventually led to the troubles in Serbia and Kosovo, but Clinton seemed to provide American involvement competently within the context of international law and the European jurisdiction of NATO. And besides, American involvement in the Balkans was far less than in the Gulf War. Troops and equipment were drawn mainly from forces which were already positioned in Europe for this sort of purpose. It seemed that the international order created in NATO and the United Nations, over such a long time at great cost, actually, finally, did work. The imperfections seemed tolerable, lessons for the next time.

Consequently, the main thing was to get on with making money, as much as possible. And so the Me-First generation arose to the challenge to stand at the corporate trough. With a permissive Mass Media, a collapsing regulatory environment which allowed corporations to merge virtually at will to eliminate competition and to restructure the once highly regulated communications and energy industries, with strong support from a Congress which was virtually controlled by a small clique of right wing ideologues under Newt Grinwich, the times were perfect for Ken Lay (the CEO who managed the most corrupt corporation of all time - Enron). And times were also perfect for the long cast of villains who are partially described in Arianna Huffington's populist broadside "*Pigs At The Trough*". Central Casting sent Ken Lay up with impeccable timing, along with a legion of one hundred thousand other greedy locusts who began to systematically corrupt American corporations into vehicles of plunder for scamming the rest of the social system with high prices and high quick turnover products to fix or solve illusionary problems like "energy shortages".

The Mass Media loved the numbers. They kept pushing the hype with very few questions right up until the break...leading the lemmings pretty much like they always do. Er, sorry, these days, the metaphor is no longer 'lemmings", it's "sheeple".

Manifest Destiny IV

It is easy to see why a great many American "Good Citizens" would have such fervor in their belief about progress and the rightness of their country, especially those who came of age during the 1980's and 1990's. Things had been incredibly privileged for a great many of them, especially those who matriculated into the corporatized scene of the 1980's and 1990's. A long Bull Run gave many of them a decent equity stake. Real estate values continued to soar during the 1990's, also giving rise to their feelings of net worth and security.

For nearly twenty years the world seemed to finally be turning in positive directions. The original idealism which had been set forth in the aftermath of the Great Depression and World War II seemed truly vindicated. Politically, many of the worst aspects of the 20th century seemed to be righting themselves, almost by magic. Tyranny was dissolving always everywhere, liberal democracy was stronger than ever, and a third of the world seemed to be entering into a broad prosperity. It is easy to see why the boomers would be so earnest in wanting to be "Good Citizens" and rise to whatever history required of them. The "Good Citizens" can be forgiven for thinking that they had uniquely invented "destiny" and that the "Utopian Dream" was their very own.

From this we can see how easy it has been, so far, for those sly foxes in the boxes of the Mass Broadcast Industry to bring us into Act IV of our old familiar play, Manifest Destiny. Coming right on cue about 20 years after the first national linkages were in place, a new Media Megaplex of internet, cable, satellite, wire, ribbon, fiber, and relay channels had mushroomed during the last 10 years to create a mass communication industry which has so many channels to fill that it is starved for meaningful content.

Then the economy began to sour, the Dot-Com Bubble burst, and corporate scandals began to proliferate. Into this sorry scene of bad news, stepped Osama Bin Laden. His timing was nearly perfect to deliver a rousing trauma and a great panic to fill endless time in the new media channels.

So began the ultimate reality show of Manifest Destiny IV, "The Empire Strikes Back". It plays wonderfully with great visual effects each and every day on multiple channels which enthusiastically repeat 24/7 its hypnotic mantras, images, and claims. It began appropriately enough with a simple and sincere determination to begin a worldwide response to get rid of terrorism, especially Al Qaeda. Destroying their bases in Afghanistan was well received by the global media. Everything about the hunt of Osama Bin Laden made sense.

But then the good old boys from darker days crashed the play and highjacked the cast. With chutzpah of truly astounding proportions, America's jingoism was refurbished to serve a neo-Zionist agenda in the guise of the American Manifest Destiny, Act IV. A small group of people, whom we have named the Imperial Faction, promoting vigorously through various forms of the mass media, led fully 50% of the American public to believe that Saddam's regime had some responsibility for the 911 attack on the Twin Towers. This same portion of the American public was further led to believe that Saddam possessed terrible chemical or biological weapons.

This created a dangerous illusion, a wild rumor of the sort that causes panics. In fact, it did create a panic. If you proclaim that Saddam gave some sort of mystical (non-specific) support to international terrorism, you can get people to suppose that he might give some of his weapons at any time to his supposed terrorist associates. This of course creates an illusion that Saddam's continued existence was an imminent danger to Americans

In the minds of many these claims and illusions fostered by the American government almost certainly must be true. Many well meaning, trusting people who want to be "Good Citizens" and patriotic Americans, just like the "Good Germans" and "Good Japanese" did some 70 years ago, want to put faith in their leaders so that great things can be accomplished. So they believe these things. They want so earnestly to do right and to believe in their leaders.

But there is one slight problem. These claims are as big a whopping big lie as any ever told by Goebbels, Hitler, or Stalin. Goebbels would have been very jealous at the success of the Imperial Faction in telling this Big Lie.

Is there anything wrong with the people who believe these fibs? No, George W. Bush told them these fibs more than once quite clearly in a variety of ways. So the main problem is elementary. A liar has implanted false ideas in their brains. They have been psychologically raped by crooks.

This is disturbing enough. It is even more disturbing that the American Mass Broadcast Media permitted George Bush and his cronies in the Imperial Faction to continue to endlessly repeat this Big Lie without calling it for what it is. They endlessly allowed old "Wartime" to continually make the claim. They falsely allowed the story to communicate that George Bush said that he had proof that Saddam is linked with the terrorists. They left the impression that this was widely accepted as a fact. They allowed this impression to replay countless times as a hypnotic suggestion which gathered more and more power as time went by.

They could just report the facts. They could report that:

> "Once again "Wartime" Bush has made a claim about Saddam's links to terrorism which about 80 percent of planet Earth believes is a Big Lie. Large numbers of people in many nations are getting very concerned about Bush's personal credibility and wonder what is wrong." They could add: "Many viewers think we have a credibility problem, what do you think? Be sure to send us email."

But of course, they didn't and they often still do not report the simple facts. They prefered to broadcast "the slant" as cast by the Imperial Faction, along with dramatic flourishes in the music. Now ask yourself, why? The answer to this question is the main problem we have in being able to manage our future.

Fortunately, some portions of the broadcast media are now beginning to suspect that they have been taken to the cleaners. Some editors are beginning to find the "Big Whoppers" of the Bush Administration and the Imperial Faction as dramatically entertaining as "The Empire Strikes Back".

The Red Scare Scam and the Goons of the John Birch Society

Perhaps it is accurate to describe the "spell casting" method as similar to the way in which a right wing bigot from Wisconsin by the name of Joe McCarthy invented the "Red Scare" of the 1950's. By exploiting the insatiable desire of the American Mass Media for sensationalist headlines, Joe McCarthy served as the poster boy for kooky right wing groups like the "John Birch Society". He created an intense fever of fear against evil communists who were said to be infiltrating America everywhere as part of a plan of Soviet (Jewish Bolsheviks) conquest.

He never failed to get headlines and TV coverage for his congressional "House Un-american Activities Committee". He would subpoena people to testify at his hearing, where he would grill famous people about their friends and associations. It was an early version of the "reality soap opera genre". With a combination of innuendoes and one-sided slanted testimonies, he managed to find a few people with socialist or communist beliefs, ruin hundreds of careers, badly scare hundreds of thousands of labor unionists and liberals, and add substantial fuel to the determination of "Good Americans" to wage the "cold war" against the "godless comintern conspiracy" to take over the world (whom Birchers down the block during the 1950's and '60's were sure to whisper were Jewish Bolsheviks in league with mongrel colored people). It of course came to no good end. McCarthy finally became the target of the press and lost all, including his life to chronic alcoholism.

It is amazing how the world keeps falling for the same old saws. The overt racism is mostly gone. But you can substitute terrorism for communism and you pretty much have in the Mass Media a modern up-to-date version of the 1950's Birch bubble mania of "Manifest Destiny". Ironically, this time many Jewish proto-fascists are leading the charge to establish American dominance over the world in order to beat back evil terrorists who are plotting to destroy the American-led World Order and Western percepts of reality. Is Central Casting playing a joke on us?

In this era, the jingo agit-prop epistles of Manifest Destiny IV are conveyed through the daily ministrations of a large number of Boomer Apostles who preach in the broadcast media. To shove the country into Iraq, they waged a relentless campaign for nearly eight months, skillfully building tension and suspense about the arms inspections programs of the United Nations into a national hysteria about the imminent dangers of Saddam Hussein's linkage to Al Qaeda and related fundamentalist terrorists.

Calling it news, the Boomers of Mass Hysteria Broadcast Media delivered what the Imperial Faction most desired, they artfully created the "appearance" of a national consensus to bypass the "irrelevant" United Nations and directly and aggressively invade another country to get rid of a despised despot. Incredibly, they made it "appear" that only a few kook Democrats opposed the action, which enabled the operation to proceed without declaring war, solely on the command of the world's new "Imperial Proconsul" who sits in the White House.

And so the Americans precipitously invaded, in the same manner which has always been typical of American Jingoism, an impoverished country with a third-rate military system which could not possibly pose a serious threat to the U.S. Military. With creative juggling of opinion polls, they even contrived to have Gallup tell us that a huge majority of Americans supported the Imperial Faction's invasion of Iraq, even as the rest of the world's media informed us that 90 percent of the world population thought the idea was nuts.

Quite obviously, if you lie the big one long enough and often enough, it works! After Iraq, there are another dozen sandy desert countries in the Middle-East, almost all of them oil rich, which the neo-zionist Ubermensch in the Imperial Faction have targeted for regime change. The Imperial Faction is ready to roll hard and fast... with internet documents, plans, papers, arguments, supporting scholars, media toots, newspapers, mega wealthy donors, think tanks, the Republican National Committee, congressmen, control of the greater part of the administrative branch of government, a President who has the stamina to keep belting out hypnotic mantras to feed a lapdog press on a daily basis, and squadrons of Tele-evangelist fundamentalist preachers who are mobilizing national prayer circles for what they are unabashedly claiming is one of the greatest Presidents of all time whose visage is destined to be carved (by them) into Mt. Rushmore.

Do you think I am exaggerating to make a point? It is impossible to exaggerate this kookiness. Already the stage is being prepared to set-up Syria and Iran as early "regimes of interest" in what "appears" to be an unrelenting campaign against terrorism and/or regimes which harbor terrorists. Rumors abound that moves against these countries may come after the next Presidential elections of 2004. Even Saudia Arabia is being set-up with a variety of reports, leaks, rumors, and toots about how flaky the House of Saud really is and how it is likely to fall to radical Islamics unless they change their ways and cooperate with the American Juggernaut which is already busy setting up a long term occupation of Iraq.

So far so good. The Boomers in the Mass Broadcast Media are happy to broadcast every hour on the hour today's latest sequel to "The Empire Strikes Back". They mindlessly repeat unproven U.S. claims and assertions about terrorist links and connections in these countries. As late as May 2003 it had not yet occurred to the Mass Broadcast Media that so far the claims of the Bush Administration about Saddam and Iraq were batting at ZERO. A generation ago, the print media already would have been at the President's throat, and in fact, some of them began soon after the invasion of Iraq ended and the quagmire began under the aegis of a completely incompetent management with no plan at all for the occupation of Iraq.

Mass Sorcery And The American Hysteria Broadcast Media

Before we can leave this account of this truly mad affair, we need to examine in a little greater depth how the Sorcerers in the Mass Hysteria Broadcast Media did their work before and during the invasion of Iraq.

Many very strange people work for very high salaries in the Mass Broadcast Media but they tell the Americans very little of what really has been happening. They distort much of what does occur around the world into stories which are presented to be a supportive "line" for whatever initiatives the White House wants to promote this week.

It is as if on cue from a central director, the major cable news and old broadcast networks turned into virtual "sorcerers" to drumbeat the mad push into an unnecessary, unneeded, unwanted invasion of Iraq. If you spend hours on endless days watching CNN channels, FOX, ABC, MSNBC, and others, one key becomes fairly obvious. The most important aspect of their work is selective exposure. They will repeat the same lies endlessly, all of them like a pack of dogs chasing the same rabbit. They will also avoid the same truths, or finally make only an offhand mention of something which contradicts "The Line". The contradiction is rarely repeated.

During the run up to the invasion of Iraq, the news programs occasionally gave the appearance of reporting stories which tended to conflict with what they were promoting. They occasionally asked questions which are potentially embarrassing to the Imperial Faction. They did give nods of distracted, brief attention to peace marches, yet through it all they kept "framing" every story segment with a strong presumption that U.S. military activity in Iraq and the Middle East was inevitable.

With artless in-your-face placards and martial theme music announcing well in advance the "*War In Iraq*" , the Boomers of Broadcast melted down and destroyed in less than one year most of the claims of journalistic credibility to which the Mass Broadcast Media had aspired throughout the previous generation. Obviously, the personnel managers and talent recruiters of the Mass Broadcast Media urgently need replacing.

The Broadcast Boomers showered their audiences with an endless parade of rumors, innuendoes. gossip from "unnamed sources", flip comments, nattering speculations, one-sided perspectives, crafted omissions, leading questions, pompous "expectations", inaccurate re-statements and quotes, often grossly incorrect extrapolations of what people had just said, and, almost daily, an astonishing tolerance and willingness to communicate, raw bigoted and racist remarks from some of the worst perennial bigots ever to appear in America, such as William Bennet, Bob Jones, Pat Robertson, and Jerry Falwell.

Occasionally, the presentation of bigoted minds, without balance and proper framing by healthy perspectives, made reporters, editors, and media networks into actual accomplices in the commission of hate crimes. The broadcast media networks became the purveyors of millions of dollars of free TV promotion for demented minds and their sicko authoritarian organizations. And they are still doing it. Tens of millions of people acquired deeper layers of disgust for American TV News this past year than they thought could exist. But the Broadcast Boomers still have not a clue.

While pretending to be just your average friendly reporter, many personalities in the American Mass Broadcast Media demonstrated to the world an amazing lack of historical perspective and depth of subject by insulting entire nations of people with flip comments like "the French are just being French". They thought nothing about being presumptive know-it-alls while driving millions of Americans to use the internet to find more objective sources of more comprehensive news. From provocation such as this and many others, one of the great casualties of this nouveau-jingo era will be the Mass Broadcast Newscasters.

Wolf Blitzer: Master Media Sorcerer Teaches Sunday School On How To Invent The News

A good example of manufacturing the news by "selection" was probably Wolf Blitzer's Sunday Late Edition on CNN on February 16, 2003. His show was then formatted to assay what was the most important news of the week and from this to provide an objective overview on the issues of the week, drawing from various knowledgeable people or from those directly involved in the events.

It was a very heavy news week but it was easy to pick out the main blows. Without a doubt, the most important news of the week was related to Iraq and the main news was the collapse of international and domestic support for an U.S. invasion of Iraq to disarm Saddam. This left the Bush administration isolated domestically and internationally.

Adding to the isolation, Bill Clinton, in a prominent release, rejected unilateralism and issued a call for working with a legitimate consensus in the U.N.

Senator Byrd, in probably the most important and grandest speech of his life, delivered a speech in the Senate lambasting the Bush administration for gross incompetence. He claimed that severe international damage was being inflicted on long term American interests and he issued a grim report to the nation and the Democrats. He warned that 30 years of bilateralism in foreign policy had utterly collapsed.

The international scene was truly dramatic. Belgium, France, Germany, and Russia rejected Bush's call for using force in Iraq and began outlining plans to significantly expand surveillance in Iraq for at least the next six months. On Saturday, peace party anti-war demonstrations in over 600 cities around the world mobilized 10 million or more marchers to reject cooperation with the U.S. Imperial Faction. Not least, Blair re-iterated at least three times that he now supported going to war in Iraq only under a U.N. Mandate.

Wolf Blitzer, without hesitation or self-doubt, virtually ignored all of these developments, providing only two very short interviews with French and German ambassadors, badgering them with presumptive questions about why their nations were hindering American efforts to disarm Saddam. He promptly ignored their answers, as if he had finished with some formal need to provide balance and objectivity. He then returned to the main theme of his two hours on the air: interviews and news clips from the cold warriors of the military industrial complex, most of whom proclaimed that the time was up and war was inevitable in a matter of a FEW WEEKS.

With this particular show, which was so at variance with the real news, set in the midst of the powerful anti-war currents, Blitzer inadvertently revealed why the Mass Broadcasters had been cheerleading the inevitability of a war in Iraq for the past several months. Their editors knew something which we did not know, and they were not willing to tell us forthrightly about it. They already knew an invasion was certain and they were on duty promoting "an atmosphere of support" for it.

In other words, it was plainly obvious that Sunday in February that the Mass Broadcast Media, as a group, using a combination of selection and omission techniques, had been lying to the American public and the world and they were not about to stop their lying despite the fact that millions of people were demonstrating in the streets of the world against the invasion.

Blitzer and his crowd mainly demonstrated that an highly self-centered "media ego bubble" on the Eastern Seaboard had turned Bush's fantasy of war against the "Axis of Evil" into a mass hallucination broadcast to large numbers of people in the space of less than one year. Reality did not matter and was not news. Only the fantasy mattered, styled in the new jingoism of Manifest Destiny IV. Rather sadly, the Blitzer broadcast exhibited the extent to which the Imperial Faction had succeeded in co-opting government and media in Washington DC and New York.

The strength and size of this shared "ego bubble" (which is strongest apparently among the yuppie class in large East Coast cities and the corporate jet-set, including a great portion of the employees of mass broadcast companies) directly demonstrates the incredible power of right wing mantras ("time is up", "war is inevitable", "Saddam is evil", etc.) which had been repeated endlessly by Bush and the Imperial Faction for the past year.

Goebbels, the Nazi architect of propaganda for the New World Order in the "New German Millennium", would be in awe. No, he would be in shock. The Mass Hysteria Broadcast Spin Shysters, most especially CNN and FOX, repeated the mantras endlessly, with great visual flourishes, martial music clips, and vast quantities of supporting filler material, ranging from large numbers of TV ads for extremely violent movies and TV shows, along with hundreds if not thousands of hours of war machine footage, troop movement news, and so on.

This programming cannot be considered to be news casting. And it is definitely not MERELY entertainment. As we can plainly see, after 20 years of progressive consolidation of national media news companies from 50 companies to only 10 major ones (some say 8), American broadcast journalism has very nearly died as a credible professional activity. Perhaps it might more appropriately be called Mass Hypnosis but that does not get close enough to the sinister aspect of this massive effort to manipulate a huge population with all the powers known and practiced by Madison Avenue Advertising Agencies.

Perhaps "Mass Sorcery" gets closer to the truth. The more one looks at the patterns in these networks, the more many of the ludicrously over-paid, self-designated media idols now easily can be seen to be as manipulative as the liars they interview. Perhaps it was when CNN adapted the slogan "The Most Trusted Name In News" that Orwellian Sorcery began there as a conscious activity. Perhaps that was the time to realize that CNN had just become the name to most distrust, along with all "news" channels, and that broadcast journalism has very nearly died as a credible professional activity.

The bias of the Broadcast Media is especially obvious where their denial can be seen and cut with a knife. For the first ten weeks of 2003, the Broadcast Media pretended that there was a big question mark about the outcome of the U.N. Security Council vote. Each day they grasped at straws as to which country might or might not veto the Bush/Blair proposed resolution. Stories were broadcast almost daily about whether or not or how in fact the U.S. might get nine votes to pass its non-war declaration of war.

But all of this was a self-created denial of what is going on, or perhaps it was an "obfuscation" of what was going on. Only the reporters and anchors of the American Mass Broadcast Media "didn't seem to know" or APPEARED "not to know". The reality, which is nearly proven by Blitzer's mid February 2003 Sunday show, is that there was no mystery about it at all for weeks, perhaps months before the "time was up".

Most of the countries had all made plain their attitudes well before February. It was clear that only four or six votes could be found in favor of the U.S. position. Even as the Broadcast Media told this "story" as a question mark, Wolf Blitzer's smug mid-February broadcast made it clear that at least some news people in the channels already knew from "sources" somewhere that the war really was "inevitable".

All of the drama of the unknown about who was going to do what next was just the War Hysteria Media's hook and crook for creating mass viewership for the latest "thriller suspense" in the play of Manifest Destiny IV. Tune in tomorrow for the next thrilling installment. More probable than not, for nearly a month the War Hysteria Sorcerers were selling the American public a completely bogus story.

The "war to eliminate weapons of mass destruction", which was reported uncountable times from this slant, was probably the biggest bogus story of all, probably the biggest bogus story of the past 30 years, probably the most egregiously cynical bogus story since Lyndon Johnson's Texas-sized fibs about the Gulf of Tonkin. No nuclear, biological, chemical, or other forbidden weapons have been found in Iraq. In the end, it was always about Bush and Saddam, not terror, and it probably is also about oil, just like a great many Eurasians will forthrightly tell you.

Doubtless MAJOR PORTIONS of the American Mass Broadcast Media have been revealed to be not just lousy news channels whom the left dismisses as lapdogs of the politicians or as pandering, corporated war whores. "Lapdogs" as a descriptor does not get nearly close enough to the dark forces at play. They are much more into the manipulative arts, working more as Sorcerers who have been busy casting an "evil hypnotic spell" toward war this past year in support of the Imperial Faction.

As one would imagine of sorcerers, one of their primary accomplishments is the cultivation of Great Presence, all the better to serve their Great Pretense of profoundly pontificating on dominion and power and the one world superpower "uber alles" (German for "over all of us", used as an hypnotic mantra by the Nazis). Unfortunately it would seem that the time spent cultivating their "presence" has come at the expense of their brains.

The Precipitous Loss Of Credibility

The great success of the American Mass Broadcast Media in creating and delivering the invasion of Iraq on George Bush's platter has brought forth much else as well. As the purveyor of such incredulous lies, the Mass Broadcast Media has brought the fatal demise of its own credibility among huge numbers of Americans and people around the world.

In retrospect, looking back at the long play of Manifest Destiny since World War II, some of the ludicrously overpaid Mass Broadcast lapdogs and war whores have been dutifully serving up American jingoism since Vietnam. But it is the issues related to what they have done with their electronic Sorcery in helping Bush which stinks and will continue to stink in the history books for a very long time. They have helped Bush violate both national and international law while diverting the international War on Terrorism" into an imperious U.S. invasion of Iraq.

How can we avoid wondering if the people who control and manage these news networks are compatible with civilized democracy and representative institutions? It seems dubious at this time in history. Some people are mounting major legal campaigns against the Bush administration to hold him and his henchmen accountable for making war illegally. They ought as well to go after the Sorcerers who engineered the climate of opinion to make it all possible. Perhaps the greater part of the current coterie of media management and editorial control in the American Mass Broadcast Industry should be summarily replaced by people who are 20 years older and have much broader historical perspective. Or any random selection of Grandmothers from North Dakota or Iowa, who doubtless have superior ethics and a much keener sense of why half the things many politicians say is phony.

Inevitably, like all bubbles, this current bubble play of Manifest Destiny IV will crash. Somewhere, somehow, something or somebody will prick the bubble and Bush with all of his men and all of his horses, along with many of the pundits in the Mass Broadcast Media, will come crashing down to Earth. Eventually they too will have to deal with reality along with everyone else. The question is, when? The story line for the media, is "how".

Perhaps the incompetence of the U.S. occupation of Afghanistan and Iraq will cause a long slow deflation as people gradually lose confidence in Bush's competence. Perhaps bellicose actions against Syria or Iran will prove too much. Perhaps stupid remarks by agents of the Imperial Faction will rebound against Bush and cause a cascade of reactions which reveal too much for the public to stand. Or perhaps organized political activity among the Democrats, the Greens, or even among outraged Conservatives in the Republican party, will cause enough damage to the bubble to dethrone the Imperial Proconsul. Or perhaps the hypnotic veil over the lies will gradually wear thin just enough to enable enough people to wake up to vote Republicans out of office.

Perhaps, just perhaps, it will be the Big Lie itself about the "weapons of mass destruction". They still can't find them as of July 1, 2003 and the mass media is getting increasingly twitchy about that. During late June and early July, finally, many news venues, even Wolf Blitzer, began to explore the question of the "weapons of mass deception". The Boomer Apostles began to question their faith in old "Wartime". They began to present the possibility that the entire war had been predicated on conscious acts of deception. Perhaps exposure of this failed Big Lie will begin unraveling the wicked webs of the Imperial Faction.

Perhaps some portions of the Broadcast Mass Media will partially redeem itself by genuinely examining and communicating what is really going on. They will find it fairly simple to do. All they have to do is take up and copy what major portions of the print media have already begun, a serious exposure of the lying ways of the Imperial

Faction and how it manipulates the press to deceive the American people.

Perhaps it will all come down to the forged 'uranium cake' documents. What did you really know Mr. President? And when did you know it? Who covered up the CIA objection? And who was he working for? The truth is, the plain simple truth is, Richard Nixon's dishonesty was a high school prank compared to the culture of disception, lying, and manipulation which has sprung up around the current Republican administration.

The Rise Of The Global Media

No one can predict how the bubble will break nor when but it is probably safe to say that the new internet-news media will play a considerable role. The American Mass Broadcast Media and the Imperial Faction accentuated and accelerated mightily the development of a world trend which is well worth examining. Intuitively, tens of millions began to see that TV News is heavily manipulated, even more so than in the past, by political propaganda. To find out the real news, millions of people began to use the internet to seek out new media sources which could provide more complete information from other perspectives.

They were not disappointed. The unedited newsfeeds from UPI, AP, ITN, and others could be easily accessed from Yahoo and many other free sites. Better yet, many new sites on the web specialized in analyzing the agit-prop slants, omissions, and selective, ahem, "storytelling" which appeared in various portion of the mass media.

As well, many major newspapers in nations around the world now have English language versions of their papers on the internet. The slower print media was still thinking, was still doing news as news, and still offered a variety of opinions and interpretations about what was happening. Collectively, the world's print media were not shy about reporting international events from diverse angles. If one looked, the American drive to Baghdad could be seen from a far less nationalistic, more disinterested, appropriately skeptical point of view.

By comparing the various approaches to the world news from various sources and nations, it was much easier to see than ever before that the headlines in the Mass Broadcast Media were hackneyed and slanted, often dumbed down for moranic levels of thinking along nationalistic "slants" .

The new "internet-media" which is gestating through the internet is generally less nationalistic and more international in perspective. Russian, French, Pakistani, Egyptian (to name just a few) perspectives on current affairs are now as readily available to citizen households as the old fashioned wire services which used to spoon feed national and international news items very carefully only to newspapers and radio/TV broadcasters.

New perspectives, new institutions, new personalities, new politics, all will flow inevitably from this shift of attention to the internet-media as a primary source of "perspective" news. These new internet "sources" and methods of communication will no doubt have a profound impact on sustaining new content for the rapidly expanding satellite/cable channels which are appearing around the world. How else are reporters and editors going to acquire content, check facts, and pull a variety of sources together for the new wave global journalism?

As Marshall McLuhan observed a few decades ago "the medium is the massage" which shapes the social and economic agendas. As this "agenda" setting power flows through a plethora of new trans-cultural media sources, all national and international politics will change. As the old fixation on a nationalistic "American" point of view breaks down, what is "news" becomes more "planetary". As the news becomes more planetary, people on all the continents begin to have a hand in setting the "agenda" of what is "in the news". And thusly many cultures are beginning to shape the common awareness around the globe of what is truly happening.

We are in fact seeing this "change" happen now more strongly and rapidly than ever before. The current Bush-induced polarization in the world is accelerating the formation of both the emergence of a world climate of opinion and the rise of the new media, both of which are reactionary against Bush's support of the Imperial Faction.

This is an incredibly healthy development. We are seeing quite clearly, for example, that the internet is the primary reason why a new climate of opinion for peace, justice, decency, and equality is being generated all around the Earth. Without the internet, it is doubtful that the peace movement would have emerged so quickly and powerfully to challenge the presumptions of the Imperial Faction.

Irony of ironies, the internet is the child of the Pentagon which engineered the system to enable the military-industrial complex to efficiently communicate and coordinate research and development activities. Now it is peace movements and open international intelligence-sharing which are waltzing around the old barriers of space and time to coordinate the emergence of planetary perspectives. Central Casting must be having a good laugh.

As can now be seen abundantly, while dot-com wannabe billionaires were generating a thousand unwanted commercial sites, the internet was busy being transformed into a huge, strong, literary, mostly non-commercial, trans-cultural community of common interest and news. The dot-com greed puked failure all over itself during the past few years but the shared information sources and communication links have blossomed luxuriantly. For every failed commercial web site, a thousand websites sprang up to share the creation of a large variety of news channels, information repositories, data analysis sites, and public interaction and transaction modalities. The "successful" part of the internet is rather quite like as the pioneers of the internet imagined and intended before scamsters tried to invent ways to "own" and "sell" cyberspace.

Many of the new internet media are heavily cause-oriented, ranging from neolithic reactionaries to radical anarchism. A large number of peace and justice movement groups can be found now appearing all around the globe in many languages.

"Internet activism" along more practical lines, for community, trade, and a great many special interest focus groups probably will continue to increase exponentially. Hundreds of thousands of "groups" are emerging which link small numbers of people in local areas to global networks along all dimensions of interest and concern at many levels of participation. The internet revolution has hardly even begun to match the creativity and desires of humanity.

As this activity blossoms, indeed even as it has already blossomed during the past year alone, the slants, selections, and omissions which make up the claims and games of national elites are being eclipsed in favor of universal values and perspectives. In this new internet "Global Village", all local news and TV network monopolies are "busted". The best efforts of dozens of reporters and news organizations can be scanned and many angles on events are easily found. Hundreds of websites offering specialized expertise or highly focused perspectives and analyses are easy to find. Many of them, like the Russian intelligence analyst's daily news on the "real" progress of the invasion of Iraq, which appeared for a short while before being shut down by unnamed pressures, could be found to be amazingly good, true treasures in a sea of American news mediocrity.

This experience was broadly shared around the world that it fundamentally changed the news habits of millions of people and directly shaped the sudden emergence of the anti-war movement. With a much longer "vision" about what was happening and how various political and media leaders were manipulating masses of people, peace activists were able to put two and two together well before the war and began the first anti-war mass movements to march before events had actually created the war. They created new corners in the internet to operate a collectively shared tool for creating peace and sharing new international perspectives for a truly world view.

Millions of people who support a strong international movement towards peace and international justice have become aware of each other's mutual presence and support around the world across all national boundaries. This awareness blossemed through direct connections to the internet and partly from access to global media sources far beyond the blinders of American editors.

The "Global Village", which Arthur C. Clarke described 40 years ago, has begun to emerge on digital wings. Stimulated by worldwide media attention to the War on Terror and the availability of prodigious quantities of information on the internet about the drive to war in Iraq, a mutual sense of a shared world climate of opinion, above and beyond nationalistic identities, is growing among Europeans, Muslims, Asians, and North Americans.

By mutually opposing the invasion of Iraq around the world, great numbers of people have experienced a permanent shift in consciousness to this global level of focus. They also have become aware of the growing movement of hundreds of millions of people into this same global consciousness which desires to make peace, not war. This shift in consciousness accelerated substantially the ability of peace and social justice movements in countries all around the globe to parallel and coordinate their activities and events. The internet permitted this newly emerging consciousness to gain strength and parallel new events and information throughout the world within a matter of hours and a few days.

This awareness and electronic linkage greatly expanded the demonstration of worldwide opposition to the Imperial Faction's plans in Iraq. We can only conclude that internet media will continue to play a major role in sustaining and fueling the growth of opposition to Manifest Destiny IV. Indeed, this same dynamic is likely to create rapid opposition to the activities of any country which undertakes an aggressive path towards violence. Self-seeking will now evermore find its old manners of rationalization and propagandizing less and less convincing in a world where a multitude of informed third parties can provide alternative points of view directly to anyone, anywhere, across all national boundaries.

One can easily suspect that five years from now analysts may argue that the crystallization of the internet into a new news medium and political coordination tool for millions of people was one of the greatest long term impacts of the Imperial Faction's jingoism, a highly positive outgrowth which helped humanity grow past an ugly era.

This phenomenon, which is doubtless as significant as the PC revolution, has barely begun. It will grow greatly and transmute in a variety of entirely unforeseeable ways. Because of the powerful "agenda setting" effects of this democratized mass consciousness, people working through the internet are highly likely to profoundly affect the economics of the next several years, the nature of the "Wall", and the process of recovery on the other side.

Accordingly, it is most likely essential to keep this long term world trend in one's consciousness as one casts about to find how to survive the coming times. The internet will likely have major impacts on the new industries which will lead the recovery and the internet will doubtless help job seekers, professionals, and investors to find new economic opportunities in the recovery industries for those who need them.

Section Five

The Global Economic Shift

Introduction

Though Edgar Cayce died in 1945, his spiritual work has endured and from this we have gained a great many things which provide important clues for our own lives. One of these clues is about a major historical economic "crunch", a deep recession or even depression, which is due in 2006 or 2007. Based on rigorous analysis of Cayce's life work and several hundred of his clairvoyant predictions about the future, all objectively documented in the "*Return of the Phoenix*", we have the certain knowledge that this "tip" has a 93% probability of being right on the money.

Accordingly, we can use this as a given: we collectively hit the economic "wall" in 2006 or 2007. The question then becomes, how bad is this "wall". How hard will we hit it? And regardless, how can we best prepare to avoid the losses while positioning ourselves to gain for a new era of growth?

Most importantly, we should also ask whether we have already hit the wall. Perhaps the breaking of the Dot-Com stock bubble in May of 2001 was the equivalent of the stock market crash in October 1929? Perhaps the long slow slide of the economy since June of 2001 is like the same long slow slide of the economy into a deepening depression which occurred for 40 months between October 1929 and March 1933? If we have already hit the wall, shouldn't we expect a long slow slide down from the early part of 2003, down steeply for the next few years until 2006, with perhaps a few short-lived small rallies between now and then?

Shouldn't we then expect to hit the bottom of the cycle in 2006 at the latest, with the DJI at between 1000 to 2000 and with many real estate values off by as much as 30-70% from their current valuations in "Silicon Valley" and many other like Boomer bubble areas? That's the ultimate Bear Scenario based on Cayce's "Wall" and it is a total no-brainer.

To answer these questions now and during the next few years we will meet a lot of Bears and Bulls, by constantly vying for attention. The Bulls, as everyone knows, are very skittish but seemingly determined to preside over another era of major growth after this embarrassing economic pause of the past two years. They prefer to believe that there is no slow slide underway. In fact, they claim that the market is already rising. By June 2003, the archetypal Bull refrain could be heard fairly loudly: get on the wagon and ride to your fortune with us!

Despite the snorting and stomping of the Bulls, the Bears are profoundly more pessimistic than they have been for a generation. North America has rarely seen so many growling Bears pointing to signs of continued economic failure. Some Bears are even seriously arguing that the U.S. economy is now slowly sinking into another major depression. They point to another major increase in unemployment during June 2003 which raised the "official" rate to a ten year high of 6.4% unemployed.

As a result of this undeniable "skidding" of the economy, many extremely wealthy and astute individuals conclude that most stock prices are still drastically over-valued, anywhere from 30 to 200 percent. Even if the economy suddenly began to improve, these Bears argue, it would take two years before corporate earnings increased enough to justify current stock prices. How then, the Bears ask, can anyone expect the stock markets "to rally" in a new Bull run anytime soon? Better to hibernate on stocks until June 2005, and then ask if stock prices are reasonably priced.

Despite the pessimism and the profound importance of the Bear questions, there are many reasons to suppose that we might see one more growth spurt during the current 24/25 year economic cycle, which began in 1983 and is not due to end until 2006 or 2007. For one thing, we are NOT seeing a downward slide in the stock market or the economy during 2003. As of June 2003, the DJI stocks are up over 10% since the beginning of the year and Bullish sentiments have been pushing the market up consistently for over three months. This "appears" to be an augery of greater economic profitablity in the corporations and the beginning signs of recovery from the past two recessionary years. Get with it, the Bulls will tell you, increases in stock prices always lead the recovery of jobs and economic expansion.

And what about the huge economic stimulous programs which are underway? The Bush administration is spending money on national defense like a fleet of drunken sailors in port after a year at sea. Surely, snort the Bulls, another round of economic expansion must come as a result of a huge demand for war products and services, on which the U.S. Congress has authorized untold hundreds of billions of dollars above and beyond past levels?

There may never be very good answers to all of these Bear and Bull questions. It may not be possible to choose decisively between the Bears and the Bulls and it may not even be necessary to choose. Reality may remain too fuzzy, indeed there are so many difficult politics and struggles in the world producing so many adverse impacts on the economy that the fuzz may just get progressively worse.

But even with all the fuzz, it may be possible to come to good enough answers to put together a reasonable economic strategy which will work for all seasons. If we can get clear enough about the dynamics which are really at work in the economy, we may be able to find strategies which can turn on a dime to accommodate both "Bull" opportunities and "Bear" downturns. With these same strategies we can begin preparing for meeting "The Wall" in 2006 and the new growth horizons which will begin to beckon us in 2008.

This section begins, then, with the broad economic factors and the trends they are producing. Within this dynamic context, the Bear and Bull arguments and scenarios are summarized. To take us all the way to 2006, a consensus "pick" is provided. Since the "consensus" waffles back and forth between the economic extremes, it is called the "Bearme-Bullme Strategy". And finally, the economic and political logic of the unfolding global shift is summarized.

The Likely Trends Through To 2006

So far, beginning especially in Section Three, a great many questions have formed up about a large number of possibilities and alternative scenarios for how global events and conditions events might work out. There are so many questions and possibilities, one can rattle around indefinitely in confusion. To clear the dust and deal with some tangible realities, here is a summary of what may be the most likely trends.

In the "Global Political Shift", we summarized the primary global political shifts which have been spawned since the advent of the Imperial Faction and its bully boy invasion of Iraq. As a result of the Bush administration's diffident, highly aggressive moves against the United Nations and sound principles of constitutional law, both large portions of the American public and an overwhelming majority of

Eurasian peoples became deeply polarized in their attitudes against the Bush Administration and American activities in the Middle East.

This polarization was radically deepened by the Bush Administration's use of highly questionable arguments, deceptive rhetorical devices, and "evidence" which often resoundingly failed to meet minimal professional standards of objectivity. The "evidence" appeared hastily "cooked" to support the desire to go to war. It did little to prove that war was a rational response to a real danger.

Many controversies over U.S. tactics engendered a precipitous loss of credibility around the world about the ethics and reliability of the U.S. government and the U.S. Mass Broadcast Media. The latter was demonstrably more interested in supporting the broadcast of the Bush Administration's dubious claims than in providing the public with real "news". The content was so threadbare and unbelievable to so many people, the Mass Broadcast Media mainly ended up demonstrating to hundreds of millions of people that it was more interested in supporting Bush's desire to go to war than to serve the public with real information about the real world.

This has set in motion profound trends which do not auger well for the American economy. Eurasian nations are gearing up to cooperate much more intensely, both politically and economically, to ward off what is now being widely discussed as American Imperialism. This will tend to reduce international cooperation with the U.S. and increase conflict in all international political arenas, most importantly in the United Nations and in the various economic cooperation councils. The U.S. economy may suffer increasing damage or drags on its performance as a result of this growing Eurasian non-cooperation

North Americans, at least those who have naively followed the mass media "slant" and the emotional sorcery of the Imperial Faction, may feel confused by the strong international opposition to the invasion of Iraq, as well they ought to. But already it seems that key Eurasian politicians are not at all confused. They are busy reformulating their strategies and arranging for new tactics in dealing with George Bush and the Imperial Faction.

In Eurasia, the leader least understood and least visible to most of the world is Russia's Putin. He is quietly emerging as one of the most effective and important leaders in the world. In April 2003, Putin was already signaling that he was taking serious involvement in resolving the crisis of North Korea, which he proposed to do multi-nationally with China. More probable than not, this issue will be resolved peacefully without much U.S. action and thereafter the U.S. will be encouraged to gradually withdraw from Korea altogether.

With Putin's active encouragement, the Eurasian peoples are being emboldened to conceive of a planetary alliance far larger than the few countries of Western Europe. The greater part of Eurasia is being actively courted to form sufficient political and economic "superpower" to outweigh the military muscle of arrogant Americans who have been tragically led by a combative, historically ignorant man far too deeply into Lake Hubris.

From this we see that the fundamental geopolitical equation on Earth has shifted and is shifting increasingly by a large amount. It is becoming apparent that the "American Century" was the 20th century and that the 21st century belongs to a post-imperial, multi-polar, post-superpower civilization. It is also apparent that the formation of this global reality will be accelerated directly as a result of the polarization which the Imperial Faction has induced with Eurasia by its clumsy tactics in claiming an exalted status as the One World Superpower.

So it is, even during this moment at the height of the American military superiority, we can see that American political influence around the world is in serious decline and that the American economic system is more vulnerable than ever to forces far beyond its control and its shores.

Equally vulnerable is its political system. Though a few politicians with immense chutzpah still extol the superiority of the American Dream and claim that America is the greatest nation there ever was, the truth is far more humble. There are now several countries with higher standards of living and prosperity than the Americans will give to each other. Tens of millions of Americans live obviously degraded, highly insecure, depressed existences which are spawning an increasing tide of angry, disconnected, disaffected people whose political expressions can and sometimes do take quite negative forms in destablizing the political system.

People in power and elites of wealth almost always over-estimate the stability of their existing political systems and they rarely understand the depth of "disaffection" which surrounds them. This is why major political change is usually accompanied by drastic measures amidst recurring violence.

American elites are no different. Some 10 to 30 percent of the population is caught up in the East Coast, Me-First, Greed-Is-Good, Yuppie Media Boomer Bubble which Bush has come to symbolize so well. They are systematically dismantling the insitutional structure of America as they continually devastate more governement programs which have been successful for decades. Doubtless, these Boomers are fully capable of sailing the "unsinkable" U.S.S. America, like the Titanic, at full speed ahead into the icebergs of mass privation and dislocation. This is a huge problem for the American political system and it is getting worse as the economy disintegrates and the lie of the Boomer version of the American "Utopian Dream" turns increasingly sour.

While the Eurasians re-arrange their coalitions and forge new initiatives among themselves, the American "Imperial Faction" will initially test the durability of its arrogance in Iraq. History, it appears, and God, it seems, are fully equal to providing the provincial American ego bubble with more than enough stress to test its limits.

Already by the end of April 2003, all social groups in Iraq were dissolving into a myriad of fractures which are too rich for North American minds to easily grasp. Calling their initial occupation a victory, the Americans are so provincially conceited that they fail to see that they have suffered an embarressing defeat by their failure to end Saddam Hussein's bloody rein of terror in Iraq and by their inability to prove any of the Imperial Faction's claims about weapons of mass destruction in Iraq.

Like Osama Bin Laden, Saddam lives on and continues to terrorize Iraq even an as the rest of the world determines the embaressing truths about old George "Texas Tall-Tales" Bush. By the force of his will and money, and that of his minions, Saddam succeded in organizing a guerilla war against the U.S. In the space of two months, Iraqi insurgent forces have been able to initiate a campaign of sabotage in the oil and energy industries while conducting random killings of Americans, Brits, and those who cooperate with them.

If you add Al Qaeda and some of the Palestinian terrorist groups to the mix, it is plain to see that these guerilla hits have become daily victories in what is aleady an emerging guerilla war stretching throughout the pan-arabic world from Palestine to Afghanistan.

The whipsaw, sensationalizing, either/or mental processes of the American Mass Broadcast Media will make these events in Iraq and the Midle East impossible to understand through the reminder of 2003 and 2004, even if the media instant experts purport to reveal all. Americans will continue to experience more Vietnam deja-vu and confusion about what to do with what is certain to be a growing mess.

The exceptional difficulty in restarting the economy for the past two years provides even more room for confusion. Despite offering extremely low interest rates and other easy monetary policies combined with tax cuts, the old economic stimulation magic has failed. So has the huge Keynesian Economic Spending Plan which has been initiated. Of course, the Republicans will never call it this but this is what they are doing. The runaway pell-mell blank check spending spree for Defense related expenditures is a vast Keynesian program.

This wartime-like mobilization already should have initiated a boom of corporate spending and economic expansion. Just the anticipation of the flow of orders should have the major corporations buzzing with high energy positive signals. The cash registers should already by ringing. Yet in June 2003, an economic expansion was nowhere to be seen in the U.S. Doesn't this demonstrate that something is seriously wrong? Has Oz fallen asleep behind his curtains?

These questions will dominate the next 18 months and perhaps much of this first decade in Century 21. Obviously something is amiss with the old economics but it is hard to find answers. Nobody in Europe and Asia seem to have any answers. Indeed, hardly any national economy works anymore in a way which national governments can lead and fix when broken. Quite possibly this same moment of Zen has come to American policy-makers.

To explain the lack of response to "traditional" economic nostrums, many analysts and historians argue that there are now serious defects in the ***structure*** of the U.S. economy. It is widely claimed that these are far more severe than in previous decades. Those who see these defects also generally argue that the problems are widespread and can be found in Germany, Japan, Argentina, Brazil, Venezuela, Italy, and many others.

For the U.S., a good candidate for the reason why the conventional magic no longer works is probably the "***structural problem***" which exports production equipment and jobs overseas, mainly to Asia. More than likely we are just now finally hearing that "***giant sucking sound***" which Perot, Buchanan, and others over tens ago tried to warn us would be the result of pushing "globalist" agendas. As the demand for goods rises as a result of increased liquidity in the economy (tax cuts, low interest rates, increased government spending, modest job expansion, overseas orders, etc.), factories in Singapore begin to hum while the unemployment rolls in Detroit go up.

If so, what this means is that no matter how much money flows through the government for military goods and expansion, it may not be possible to revive the U.S. economy into another sustained dynamic growth curve. Instead, the best which may be achieved is some very modest increases (1 to 2 percent) which just barely move past the prior year's performance.

Under this condition, joblessness will be a large, persistant, nagging problem and the fiscal crisis in all local government will engender a climate of continual negativity. During 2004, this will produce a lot of civic friction, organizational fireworks, and major shifts in the political parties. Add this normal political strife to the extreme political polarization over issues of credibility which the Imperial Faction has engendered, more likely than not an intense civil war will break out within each party.

The Mass Broadcast Boomers were still propagandistically extolling Bush as late as June 2003 as some very popular president but nothing could be further from the truth. Traditional Republican "heartland" progressive conservatives (people who identify with Barry Goldwater, Nixon, Reagan, and like) are hopping mad at the Imperial Faction. They will fight loudly and they could "bolt" to a third party in large enough numbers to prevent Bush from being re-elected. Similarly, traditional liberal democrats are hopping mad at both George Bush and the faux-centrist democrats who currently control the bulk of the democratic party. Rather than vote for the faux-centrists, whom the left dismiss as 'Republicrats", major portions of the democratic rank and file may bolt to another "third party".

Quite obviously, this could create a field of four minority parties and widely divergent possibilities for all investors and job seekers. If the economy is emerging from recession, the more the Democrats will break up and the higher the stocks will go. But the more Bush's grasp is threatened by economic non-performance and thunder from conservatives, the lower the stocks will go.

If the economic situation remains unclear and confusing, the voters will probably vote against Bush in such large numbers that any of the other three factions could win, depending on the names and the platforms and the money. This condition may emerge by summer of next year.

Thus, after an initial start this year from "wartime stimulation", political factors next year may cause the economy to wallow and create such economic confusion that the stock markets stay depressed at levels not much higher than in July 2003. If so, it may not be until Spring of 2005 before the stock markets begin to show a rebound. Such a rebound will be fairly short-lived, not much more than 12 months. The economy will simply have run out of time in Cayce's *"25 Year Economic Depression Cycle*".

China Et All Trade –That Giant Sucking Sound

One of the most important over-riding dynamics in the world economy is "***the giant sucking sound***" which can be heard in the American economy, even from far off. Economists and public essayists talk politely and abstractly about the "***structural problems***" in the U.S. economy and bemoan one or more crisis which these are sure to cause.

Critics have come to point out a multitude of leaks in the flow of goods and services through the North American economy which are sinking the economy. More and more alarms have been rung for nearly 15 years now but the holes just keep getting bigger and bigger as larger and larger sums of money flow through them.

Just what is this about? What is one of these "structural problems". Let us get very explicit. Here is what economists mean when they talk like that: a great proportion of any increase in U.S. purchases of consumer or even of war products may hemorrhage out immediately to places like China rather than driving production lines and payrolls in cities like Detroit and Los Angeles.

Consider then this huge dilemma, the Bush Administration is spending untold amounts of money for war goods to fight the war on Terrorism and the invasion of Iraq with a virtual blank check given to him by the Republican-controlled Senate. But instead of stimulating American economic activity through new jobs for new homeland production, a good portion of the huge wartime increase in the Pentagon's tax dollars will be spent overseas. The money will tunnel through American corporations and end up mostly stimulating, for example, China's economy.

There are many gaping holes in the U.S. economy but the really deadly one is the job drain. Forget about other issues. Focus here. Get it: huge numbers of high paying skilled jobs in North America are disappearing. It used to be just textile factory jobs. Now its software engineers. And everything in between.

Here is what is happening. U.S. companies are moving the expensive parts of their operations abroad at an increasingly rapid rate. Production of goods AND services are moving anywhere except into depressed areas of the U.S. This reduces U.S. employment, benefits, pay, gross national product, technology infrastructure, middle class, social security income, government revenues, and the general rate of economic growth. It increases the balance of payments problem and the pressures on the value of the dollar. The value of the dollar declines, foreign goods increase in value, inflation increases, the middle class in North America joins the welfare rolls.

This process is well under way. We have just started into the decline of the value of the dollar. Accordingly, the main structural issue in the U.S. economy are the corporate rats who are jumping ship. They call it globalization and say it enhances everyone's interests and well-being. How very strange, then, that the economy of America is sinking as they globalize?

This situation has been worsening since the rise of Japan during the 1970's and has became a grave problem with China for the past decade. Ironically, Japan first went through the entire process of exporting jobs during the 1980's and since then has barely been able to keep itself from plunging into the abyss. It has frozen its economy to avoid the abyss and "pretends". How long it can continue to "pretend" is anyone's guess.

Here is what happened in Japan. After World War II they gradually built their economy with the best that money could buy to be able to excel in all technology industries. They were phenomenally successful and began to generate such huge mountains of dollars from their exports that it became a much talked about "problem". What to do with their riches which would equalize world trade?

Rather than open their economy to permit their people to import larger quantities of U.S. and other foreign goods, the Japanese corporations decided to invest their money abroad in other countries to make more of their corporated goods in the countries which were buying them. Thus Toyota came to America, and so did a lot of other companies. And they went to Singapore, Taiwan, Thailand, Hong Kong, and even China. At the time, this "felt" like success and really good politics.

The Japanese corporations financed a lot of this activity with ultra low-cost loans provided by their banks, much lower rates than charged by most banks around the world. Japanese companies put up as collateral their real estate in Japan. If you had real estate in Japan and were socially connected to the corporated political old boy network, you could get loans at very low rates and invest the capital abroad. Within a dozen years, Japanese real estate values soared to absurd levels. Everybody who wanted loans wanted to outbid all others by buying up as much real estate as possible.

This process was well advanced by the end of the Reagan years. But all "free rides" come to an end. A recession rolling through the world from country to country caught up the Japanese economy and then the American economy a little later. The wide-open American economy, based on the petrodollar, recovered fairly quickly and blossomed into the Clinton Feel-Good Bubble. The "closed" Japanese economy never did. An increase in demand in other countries for Japanese products (which were no longer made in Japan) failed to produce an increase in exports and renewed prosperity in Japan.

This caused everyone in Japan to realize that Japanese stocks were way over-valued but not nearly as over-valued as Japanese real estate. Both crashed mightily and both crashes wiped out a huge portion of the wealth for which Japanese workers thought they had been working for the past 30 years. They have never regained most of it, some 12 years later, and they are still limping along with a "frozen" economy which barely expands by around a percent or two per year. Worldwide inflation probably eats the value of that away.

To fix the problem, the Japanese government tries to stimulate the Japanese economy in the same way the U.S. tries to stimulate its economy, by lowering interest rates. But it has not worked in Japan because Japanese corporations cannot change their mental habits. They borrow the money at the current time at virtually NO INTEREST COST. That's FREE MONEY which they collateralize with foreign assets, such as factories in China, and then they spend the money building more factories in China. Taiwan, which followed Japan's ascent, is now doing the same, following Japan's descent into this strange quagmire.

Actually, it is not strange at all, it is very logical and it is easy to fix. The Japanese people need to wake up and realize that their corporate elite is essentially free-loading off their government. They need to forthrightly bankrupt their economy, take it back in receivership at bargain basement level prices, fire the current corporate elite, remove these incompetent bozos from any further managerial involvement in their society, and then restructure their corporations and economy to enable the average person to become prosperous again.

Of course, everyone in Japan who is still pretending that they are wealthy and deserve to be a part of the ultra-elite is utterly opposed to such a program. And THAT is the structural problem. The rest of it is all economist academic mumbo-jumbo dancing around the point trying to pretend IT isn't there.

It appears that the U.S. is well down the same road. Currently, American corporations are investing some $50 billion per year in China to build brand new factories to produce the goods which the Dollar Stores and Wallie World are selling through out North America. These imports from China, which are priced so low that nobody in the Western Hemisphere, including Mexico, can match, add something like $100 billion per year to the American trade deficit.

Cheap plastic toys are one thing. But now the crisis is broadening from production jobs to professional jobs. Ironically, the internet and telecommunications revolution is now accelerating this shift. At least one architectural firm in New York produces it fabrication and construction plans in Shanghai. Microsoft has built a computer research center in Beijing and it is about to slash the costs of its software technical support telephone services by moving them abroad to be manned in Bharati (India). In doing this, Microsoft will merely be copying what other software companies are already doing in India. Even telemarketing companies are copying the flight to India by setting up their phone sales operations in various Indian cities.

Looking at the trend lines of this job transfer, Forrester Research estimates that by 2015 more than three million white-collar jobs and nearly $140 billion a year in wages may shift from the United States to other nations. This sort of "flight" would pretty much finish off the economy of North America. The only thing which is resoundingly clear when facing this squarely is that none of this is in the economic nor political self interest of the average American citizen. Corporate interests are obviously now widely divergent from citizen interests in the U.S.

The corporate refrain is that these companies have no other choice. The mantra is that everyone else is doing it and so they must also. If they don't they will be at a cost disadvantage which will put them out of business. But the corporate solution bankrupts America. Accordingly, a structural solution must come to the fore, probably on the winds of strong populist political storms, to reduce the flood of cheap imports and provide a combination of incentives to invest in new jobs in North America with disincentives to invest in new jobs in other countries. Somewhere between the globalists and the nationalists there is a better balance.

A better balance is not exactly popular in this age of exaggerated extremes and the great hurrah of unregulated John Galt capitalism for the unabashed greedy. Advocating capital and trade barriers is highly socially unacceptable. For the past 20 years it has put you solidly in the anti-globalism camp, just barely one rung above racism, one rung below sexism, and just a hop and a skip from Greens and other radical global anarchy groups.

So for the time being there is probably no fix in the works until the great invisible hand of the free market arranges its inevitable solution in the form of a devastating economic collapse into another Great Depression. At that time, agendas everywhere will shift, corporate agendas will have little credibility, and the next New Deal or New National Economic Industrial Policy will be item number one.

If it goes this far, the essential generational or cyclical requirement for fixing the economy will be to get rid of the buffoons who created the conditions. This means insisting on bankrupting out the current corporate elites from all control and management. Corporations MUST be allowed to fold so that their resources can be placed into new venues under new management with very different perspectives and policies.

That is about as Bearish a policy as one can contrive but it will do the trick. Anything less is a waste of time and resources. At that time it will be possible to modernize U.S. factories and lay in the start-up of a new set of industries for the next 25 year cycle. Foremost in that new economic strategy will be a profound displacement of the use of oil and another generational round of strict, very strict regulation of American corporations, how they are governed, who gets selected to manage them and how, and what kind of levels of compensation they offer to create a just and equitable society. That would, of course, open the way to a genuine recovery and then a new economic boom very much in the same general manner as the New Deal managed for people during the 1930's through to 1980.

The Bears

There is a powerful Bear argument about stock prices which individual investors need to reflect on long and hard whenever they get the urge to buy stocks. Many professional investors, who buy stocks for joint investment purposes, such as Warren Buffet (a multi-billionare who is widely considered to be the second most wealthy man in the United States) can be heard opining quite loudly that most stock prices are still substantially overpriced. They point out that most stocks are still selling for 25 or more times their nominal per share earnings. They believe that serious investment grade stocks should generally be priced at about one third to one half less than that, down to price-to-earnings multiples in the range of 10 to 15, which is the long term historical norm.

From this perspective, stocks priced higher than 20 times earnings will probably decline in value or not grow until the company expands substantially. And, not least, investors with this perspective tend strongly now to reject companies who are not earning profits. The dot-com company, with nothing but expenses, is no longer regarded as a real business entitled to have real stock openly sold in the public markets.

This Bearish advice most likely can serve as a valuable guide during the next few years. For one thing, it tells you how the biggest buyers will be making their buys. Understanding this key point will tell you when you can expect a REAL Bull Run from the combined buying power of the big buyers of large quantities of stock, such as mutual funds, pension funds, professional investors, etc.

Accordingly, during 2003, one should not expect to see major growth results nor Bull inflation in the over-priced high tech stocks, the beleaguered airline stocks, nor most of the others which have insufficient earnings per share. In other words, do not not expect the stock market to increase by much. Consider yourself quite lucky if you see gains of 10 to 15 percent in some select stock over the course of the next 24 months.

According to the Bears, the only thing which will justify the beginning of a major upwelling in equity values will be a major increase in economic activity, jobs, and profits in a sustained growth pattern sufficient to increase average corporate earnings per share by large numbers (up 10 to 30 percent). This is not likely to occur in anything less than 2 years, thus, according to a great many Bears, any sustained "run" up of values this year is just hype, doomed to collapse and stagnation next month or next year.

From this it is easy to see we must take care NOT TO ASSUME ANYTHING. We still don't know technically how well the economy will recover this year and we definitely know that the politicians are capable of blowing economic expansion. Since we have an abundance of evidence to be seriously Bearish on the intellectual capabilities of the Bush Administration, we should not hold high expectations about the inevitability of rational economic growth which will lift our boats and pay-off our speculative bets on the economy.

Even for leaders with strong economic credentials and intellectual acumen, attempting to manage growth in the global economy is very difficult. There are simply too many variables and independent actors. As we see over and over again, politicians have remarkable abilities to contort themselves into positions from whence they shoot themselves in the back while thinking they are slaying their opponents.

All the analytic models for fixing an economy make you look at an economy as an island in a vast sea. So we talk about the American economy versus the Japanese economy. But these are merely useless illusions. No economy is an island, they are all interlinked in multi-dimensional ways which make the old fashioned ways of linear projection a useless exercise in deaf and blind dumbness.

In understanding this truth, Japan's example may be exceptionally useful. Japan, once the world's greatest engine of growth, has failed to restart its economy despite 10 years of trying. While the U.S. soared during the Clinton years, the Japanese economy barely kept even with itself from year to year. If the Japanese cannot figure it out, who can? Does it not appear that there are no simple ways left to restart or manage economic growth? Doesn't Japan's example make all linear projection of "stimulous control" and "economic growth" highly suspect.

Some economists are now arguing that the U.S. has fallen into the same structural problem which has destroyed Japan's engine of growth. They argue that the U.S. is now spiraling down into the same inability to deal rationally with its debt and the price of this load will be perpetual stagnation in the economy.

Looking at the current 2 year recession in the U.S., who can say they are wrong? It is certainly true that decreasing credit rates for corporations or lowering tax rates for individuals, the favorite tactic of American politicians, have not worked for the past two years in the U.S. Given the hugeness of the world and the manyfold agendas of its billions of people, it should not be surprising that no one grasps or controls the full dynamics of our world economy.

So let's take an even more pessimistic risk-averse attitude. We can go deeper and argue, as far as professional economists are concerned, that the unpredicted long expansion and bull run during the 1990's was largely a mystical experience with the totally great unknowable. No economist predicted it. Nor did they explain it. Nor did they foresee the end of it. Nor could they estimate the depth of the fall, etc. They cannot predict how long the recession will last and at the moment they have no idea what it will take to cause a recovery of expansion. So ask yourself, are the professional economists any good at all for getting advice for planning your economic future? And, just what is a professional economist anyway? Just who was it who said they could do anything for you?

Truth is, economic realities in the global ocean move in increasingly mysterious ways and we really do not have, never have had, much in the way of a real science of economics. At best, we can see with reasonable clarity at the beginning of Century 21 that all politics related to economics in the so-called post-industrial world, to re-use terms from way back, is pretty much on the same level as "voodoo". So, this great technological civilization poised on the brink of the development of space, still uses mainly "voodoo economics" to cobble and wobble and limp its way along from decade to decade.

From this mystical experience, we do gain a great gem of wisdom. We KNOW that we KNOW very little at all. Accordingly, no one can "predict the economy" in solely scientific material terms and this very real fact makes Bull economic predictions inherently risky. Read this paragraph again, this is a terribly important point with which you have to be solid. No positive prediction is strictly materially rational or logical, real predictions (about human beings) flow from another level of beingness, which is why economists, as a matter of science, cannot predict the human future.

Craps, Anyone?

In truth and in sum, there are many highly astute Bears and Bulls prowling the streets of the world and each position has a lot of logic behind it. It is an interesting time with plenty of ammunition for any argument, to say the least!

A few believe that an economic collapse is imminent, no later than the end of 2003. A few countries are already hopelessly mired in depression and from this some believe that a major economic meltdown, even a worldwide depression, could engulf us all by the middle of 2004.

Others, especially those who tend to be professionally connected with the stock markets, keep arguing that the economy and the stock markets are nearly primed to climb again and may do so next month (sometime soon). Many opinions for both a rise and a fall can be acquired from various sources.

Accordingly, it is very difficult if not impossible to determine whose logic and dates are the best. The struggle between the Bears and the Bulls appears to be more like a game of craps than a science. We have "come" and "don't come" bets all along the date line for the next couple of years, at least.

If we go with the predominent Bear crowd and place a "don't come" craps bet, the coming depression could begin with a true contraction of the economy in 2005 or so and this contraction wave would continue through to 2006, when it bottoms out to match up exactly with Cayce's "economic depression cycle". Accordingly, both 2005 and 2006 as contraction years could be simultaneously true to fulfill the 24/25 year "depression cycle". Or 2006 and 2007. Or 2004 through to 2007.

So what can we do with this? Know that we are highly uncertain about the future. Knowing our ignorance provides a powerful knowledge. The truth is, a great many of both the Bears and Bulls on the scene today could be roughly in accord with the truth of what will happen, just a little rough and "guessey" on the timing. They may all be tuning into the facts, but their sense of timing may be off by a year or two or three. Looking on the scene from a large historical point of view, their disagreements may actually be very small differences of opinion.

From this perspective, the best knowledge tells us to avoid the orthodoxies and the fads which support either the Bulls or the Bears. To avoid the orthodoxies, radically discount the mass media. To avoid the fads, examine the touts and tipsters and brokers with EXTREME skepticism. Hang loose, be VERY risk adverse, be prepared to play your hand either way.

The Bearme-Bullme Strategy

Can we be more aggressive in dealing with this time of uncertainty? Perhaps. Consider the Bearme-Bullme Strategy, named in honor of Dr. Doolittle's famous beast, the Pushme-Pullme, which was something like a wildebeest but with two heads, four legs, and no hindquarters. To make it as perfectly clear as economics, each end was actually an head. In other words, take a feather from both heads and be ready for action from either end. That means, be ready to pull and be pushed forward, or be ready to push backwards and be pulled when the hindmost takes over.

In a time of uncertainty and vast change, the best strategy for the lowest overall risk is one which is equally good for both of the extreme Bear and Bull scenarios. In the case of the Bear scenario, we expect a continuation of a long slide from 2003 onward into a real depression to bottom out no later than 2007. In the case of the Bull scenario, we expect a modest economic recovery beginning in 2003 to stimulate a relatively short bull run (24-30 months) before the 25 year collapse.

By closely paying attention to signals which can reveal which scenario is unfolding, we can proceed with relatively little risk of being blindsided. To play this strategy well, the key decision is to make no assumptions about who is right or wrong, the Bears or the Bulls. Suppose instead that they are both correct and both wrong and it is just a matter of timing.

In April 2003, the economy still looked as if it were sliding downward even while the stock market seemed to be inching upwards. One could still argue that Bears still ruled. But in May 2003 the economic countryside seemed to be greening in the Spring rain, the bottom of the recesssion appeared to have turned and thus the Bulls were likely to appear soon to forage on the fresh grass. Certainly confidence slowly began to rise and the DJI began to move up even as the Bush dollar began to move down in value.

From this one could argue that the Bears were in slow retreat as the international uncertainties over the Middle East waned away. April and May certainly saw Allen Greenspan, Chairman of the Federal Reserve, evangelizing this idea. But as of the end of June 2003, conditions were really all still very tenuous. The jobless rate was at 6.4%, and the Federal Reserve saw fit to drop its short term bank lending rate for the 13th time in the course of the last two years. Obviously, the bankers thought things rather more dicey than rosey. How to play the strategy from this point?

To begin with, do nothing in actuality other than study what moves you are going to want to make. The Summer of 2003 may still be too unsettled, likely to flow with contrarian impulses, political controversies, chaotic events in the Middle East, international debates and seesaws, terrorist attacks, and so on. Spend Summer settling on what you want for your investment portfolio for either of the Bear and the Bull scenarios, but do nothing with your portfolio until the Fall of 2003.

At the end of Summer 2003, and for the rest 2003 and beyond, read the signals intensely and see whether it is a Bear slide into the ground or a Bull run inching past 9,500 and beyond. To play this game, one needs to ground oneself to the current realities to watch the shifting international political dynamics and media slants which are now shaping the world. As the international political dynamics unfold, the North American and world economies will be affected, either stimulated to expand or left limp and losing ground. One should be able to catch the drift and adjust strategy as appropriate.

As of June 1, 2003, the extreme polarization which the Imperial Faction has induced into the world, especially vis a vis Eurasia, PORTENDS a difficult period for the next three to six months in international affairs. During the last week of April, President Putin of Russia went out of his way to disagree with British Prime Minister Blair and virtually insulted him publicly at a joint meeting in St. Petersburg, Russia. Nations will not agree readily about anything and the choppy, cold political weather will depress economic confidence.

This is the inevitable downside of the unilateral invasion of Iraq and there does not appear to be any Dr. Feelgood in the Bush Administration who can schmooze over the deep differences of international opinions and egos. This polarization of national egos could easily grow and interfere with rational policies and normal economic rebound. Watch these portents very carefully.

Equally as important is the continued success and strength of the Manifest Destiny Bubble which the Mass Broadcast Media has kept on steroids for the past two years. Is the hype and manipulation still successful? Or is political sniping of Bush's Big Lies beginning to put enough holes in the Bubble that neither Fox nor CNN can keep air in it? As the Bubble deflates, public and international confidence will begin to erode leaving all markets increasingly chaotic.

One of the biggest portents of deflation in the Bubble may be turmoil and turn-over among major managers in the Bush

Administration's staff and the executive branch of government. In this light, think about the announcements during the third week in May 2003 of the resignations of Tommy Franks, the commander of the invasion of Iraq, Arie Fleisher, Bush's press secretary and consummate spin meister, and Christie Whitman, EPA Administrator. This may just be "coincidence". It may also be a real manifestation of acute personal stress and unhappiness with what is going on in and around the Bush Administration.

Accordingly, keep a watch for more such signs of staff turnover. The Bubble will deflate rapidly under such circumstances. In the case of Franks, coming at this time, or some other major military figure, a resignation could be a strong sign of estrangement from a political and intellectual environment which seems incapable of generating an occupation plan for Afghanistan and Iraq which will realistically control the ground which U.S. troops have captured. If so, conditions may continue to deteriorate in Iraq and eventually erupt in some sort of new crisis. A new crisis will heavily impact the skittish stock markets.

The Modest Bull Bubble Of 2003-2006

Despite the difficulties and insecurities, let's suppose explicitly, with some important provisos and contingencies, that a modest Bull Run will begin in June 2003 and show enough positive signs in a few months to ignite a steady upswing in expectations. The coming upwelling in the economy, if we have one, is likely to quite modest, perhaps showing fits, starts, and inadequacies during the remainder of 2003 and perhaps for much of 2004, and even possibly well into 2005.

This upwelling will no doubt come as a result of several stimulous waves, some accidental, some a matter of policy. The return to cheap fuel prices, mega-billion dollar contracts to restore Iraq, paid largely by Iraq's oil production, vast new Pentagon war on terror contracts, ultra low interest rates, a cheaper dollar, and other many factors will buoy the economy this year. As a result of these stimulous waves, the North American economy may begin in the Summer of 2003 to expand modestly and slowly again for another two or three years.

The emphasis must be on the word "MAY" begin to expand. In addition to the huge increases for military expenditures, the U.S. Congress also inexplicably passed legislation which allows defense contractors to bypass "Buy America" rules in certain ways to purchase goods and services internationally. Such legislation, in the midst of a recession, is something akin to shooting yourself in the foot. The stimulus of increased spending will doubtless be diluted as it leaks into other countries.

Even so, some expansion of cashlow in the U.S. is more likely than not and some of that will probably end up in at least a slight increase in jobs. Keep in mind that the Bush administration may not be very competent in spending the money, but it is not a bunch of Hooverites debating why intervention in the economy is not a good idea. They will spend and if necessary they will spend huge amounts of dollars to stimulate the economy.

The current Republican elite, dominated by the Imperial Faction, may sometimes present the image of being for as little government spending as possible. But this is merely a poise for various purposes of tap-dancing on the foreheads of their constituency. Conservative of the use of government these people are not. The massive political and military interventions which the Imperial Faction has undertaken demonstrates no lack of doing what they think they have to do to shove reality to move in their directions. If they need to, and they probably need to, the Imperial Faction will rapidly escalate additional economic intervention.

Since ultra-low interest rates have failed so far, and since there is almost no support among economists for the idea that the tax cuts of the Bush Administration will promote economic growth, most likely they will need to do more. This will likely take the form of manipulating international currency, trade, and commodity values, perhaps attended by an even greater expansion of war activity. All of these methods have been used extensively by imperial-minded elites during the past several hundred years.

One of the easiest and most effective methods is the manipulation of exchange rates for international currency. Thus, some of the Imperial Faction may already be setting in motion a controlled "cheapening" of the dollar, which means they will support buying euros to raise its value and reduce the relative value of the dollar. That means what it sounds like, that the Federal banks will continually sell the dollar to Europeans, or anyone with a euro, at cheaper prices.

The New York Times reported in April 2003 that some U.S. bureaucrats deliberately planted rumors of this policy in Europe. Since those rumors began to circulate in mid-April, the price of the euro increased 4%. This will modestly increase the price of European goods in the U.S. and Asia.

Theoretically, this will provide benefits to the U.S. economy. The theoreticians claim that this will decrease European exports, increase American exports, make Asian goods a little more expensive, reduce the

U.S. Balance of Payments gap, and thus induce a new round of growth in the U.S. economy. If you need even MORE growth, extol the theorists, reduce the value of the dollar even a bit more.

They seemed like a shrewd unilateral American policy but it soon was waffled to respond to growing European concern. At the European Union meeting during the beginning of June 2003, Bush made a major point of informing European leaders that he supported a "strong dollar', not a weak one. Since then, the value of the dollar has held firm. In fact, it inceased slightly by the beginning of July 2003. Is this now the real U.S. policy? Did portions of the Imperial Faction get over-ruled? Or is the game more complex than we might know? We will have to watch this very carefully. This is definitely an arena controlled by hidden "cigar filled rooms" and do not expect that anything is straightforward with the control of international currency.

Now here is where such a policy could get very clever. If the U.S. government can reduce the value of the dollar while actually reducing the dollar cost of crude oil, policy makers will be able to keep overall inflation (from the import of now higher priced European goods) at zero.

Some might say that is impossible. But nothing is impossible if you own 1000 Abrams tanks and you have them parked in the world's second richest oil field. By expanding the production of cheap Iraqi oil to flood the oil markets, overall inflation can be kept at zero, the Federal Reserve can maintain ultra-low interest rates and the stimulous force on the American economy can be kept on high levels.

If these factors (interest rates, price of the dollar in euros, and the cost of oil) can be kept controlled and in balance, some economic growth may successfully be ignited by the demand which results from the war stimulous. Keep in mind that this is THEORY. Reality is often incredibly messy, especially in economics, very tricky, even contradictory.

But as you can see, the odds appear reasonably good as of July 2003. It appears that one can rationally expect growth to result from the current combination of low interest rates, cheap dollars, rapidly dropping oil prices, huge orders to rebuild Iraq which will pump large amounts of cash into American corporations, and the likelihood of a major expansion of international oil supply from the modernization of Iraq's oil fields.

One can reasonably expect that all this will probably fuel higher corporate profits, a larger demand for equities, an uptick in overall confidence, and thus another increase or "bull run" in stock values for at least a couple of years. If so, we could see a modest Bull Run provide a

10 to 20 per cent increase in the value of some equities before "the wall" is hit in 2006/2007.

With such a slow, steady upswing, most likely with several sudden short mood shifts to keep people guessing, the DJI will eventually reach the 9500 mark between the start of Fall 2003 and just after the beginning of 2004. After the inevitable winter adjustment, it may once again climb during 2004 into the 10,000 range and during 2005 it may begin to climb towards 11,000.

Higher than 11,000 it is not likely to go. There are still too many inflated stock prices and big heavy investors are going to be "bubble adverse". Just going to 11,000 will push many buttons and encourage a lot of people to $ell $ell $ell so they can take profits on stocks they bought at a premium a few years ago. So the DJI may not stray much beyond the mid 10,000's. At minimum, expect a lot of choppy mood swings of 300 to 400 points at a time with the DJI Averages in the 10,000 – 11,000 range.

Above all, don't get fooled by major rallies. Fast rebounds will be too fast. A fundamental shift in the economy has to be felt before stocks can truly begin to ride the next Bull. For the remainder of 2003, there are strong odds that current equity levels will not improve by much past 9500, if that. Beyond 9500, the market cannot remain until a real turn around is evident in the U.S. economy. So far we are empty-handed in this department, but signs seem good as of July 2003, thus a surge past 9500 may be possible by Fall of 2003.

There are some major sources of serious trouble for our modest bull scenario and there are many variations on how it could play out. Here below are some of the major factors which could dramatically alter all outcomes.

Problem #1: Political Incompetence

Political incompetence is always a possibility for crashing the party. The track record shows that the possibility is stronger than ever. In this era, everything in American and international politics is over-reactive and this tends to bring out more of the stupid than the smart. The chaos from this is inherently unpredictable.

One could argue that the churlish rejection of the U.N. and the invasion of Iraq created the confusion which has delayed economic recovery for several months. One can also argue that the deceptions over the invasion of Iraq will lead to a serious effort to impeach George Bush Junior. That could keep the economy wallowing indecisively for another year.

If more war talk and strange controversies appear to allow chaos to rear its ugly head during the next six months, this will create declines in the DJI and upsetting economic news from abroad. If so, wait no more than six months for sanity to return and the DJI to be higher than 9000. If not, convert your play to the Bear strategy. Liquidate for the coming Great Depression and plan on waiting it out for several years.

Problem #2: Osama Bin Laden

Osama Bin Laden and Saddam have both escaped. This is the most difficult scenario to deal with because there is a wide variety of things they could do and the consequences could be great or insignificant. Or maybe somebody will catch either or both of them.

Saddam is both old enough and crafty enough, he might take the smart exit and be content to hide somewhere like a Troll under the bridge with his loot. There is not much point for that old Troll to invest in waiting for a return of power. But in addition to crafty, Saddam is about as stubborn and dense as ever there was. He may support waging a long guerilla campaign in Iraq to sour everything for everybody. This will severely affect international and national public confidence, stimulate major political opposition in the U.S., and depress both the economy and the stock markets.

Osama Bin Laden is still young enough to be eager to strike a second blow, especially since he was so phenomenally successful with the first one. This blow could be a doozy, they are certainly spending a lot of time regrouping for it. Or, maybe Bin Laden really does have diabetes and no longer gives a damn about saving the Arabs from the Great Satan.

If Bin Laden strikes and it creates the same kind of tumult as the attack on the Twin Towers, take your losses about 90 days after the attack (the markets will recover to some extent after an initial sharp plunge) and activate the Bear Strategy. If it is simply the reappearance of night club bombings and related random bombings, it may not mean much after a month. Hold until you figure it out one way or the other.

Problem #3: Disease Hysteria:

Hysteria over disease is entirely possible and such hysteria could severely damage many economies, including the American economy. This would of course trash stock market valuations around the world. A prime example of the effect of such hysteria was offered by the outbreak of SARS.

The exceptional responses to SARS may have been more in the nature of a destructive hysteria than an appropriate reaction based on sound science and health procedures. As of this writing it is nearly impossible to tell from outside the health profession.

Despite the concern, regular pneumonia is killing far more people every week than SARS did. This situation could of course change and we may still be in for a rapidly spreading epidemic. It is likely to reappear in the Fall, or next year, or the following year, and it may have mutated by then into an even more contagious agent. Who knows?

Other such epidemics could suddenly arise like SARS and spread even more rapidly into new areas to produce even worse panic spirals. From these panics, economic damage could mount and a depressionary wave in the world economy would likely spread out from the orignial source area and cause other economies, even the U.S. economy, to slip into a recessionary slide. If so, ALL BETS are off. A nasty panic-driven downward economic spiral could continue and induce another huge depressionary wave.

Watch this very very carefully. Possibly the biggest wildcard IS the SARS virus. If panic over SARS or some other new disease, grows worse over a short period of time, and then worse again, over a period of three months, examine the impact on the stock markets. Is the DJI failing to move up? Is Lou Dobbs and other financial journalists and analysts describing how the new epidemic is affecting the economy? If so, don't wait too long, move to the Bear Strategy.

The SARS virus is not currently a real threat, BUT, the REACTION to SARS did spiral out of control during April and May 2003. For a few weeks, the World Health Organization advised people to cancel trips to Beijing, Hong Kong, Shanxi and Guangdong Provinces of China. Lufthansa Airlines, Germany's main airline company, cancelled direct flights between Europe and Hong Kong. The tourist industry reportedly collapsed in China even as the American aviation industry began to collapse into nearly universal bankruptcy.

The greatest impact of SARS, or some other epidemic, may be simply the hysteria and disruption it causes in various areas, such as a case of cold sniffles causing the quarantine of an airliner and a school. From this, martial law is always a possibility, ANYWHERE, and could grow to impact major metropolitan areas with nearly complete economic shutdowns for a few days to a few weeks.

The extreme reactions in this Sping 2003, to what is still a low probability health threat, show just how fragile our economic world order really is. This reaction spiral is likely to grow even larger if SARS or some other bug does jump the fences and become truly epidemic. If so, it is quite obvious that the resulting disruption could induce an economic collapse which is catastrophic.

The economic collapse after World War I and the infamous influenza epidemic which killed millions of people comes to mind. In some countries, most organized activity virtually dissolved away and it took several years for organized communal activity to return. It was in that period that both Roman Fascism and Bolshevik Communism flowered and ultimately conspired to collide in World War II. Look at the economic statistics for Europe for the period 1917-1926. They are pretty grim.

Problem #4: Money Hysteria

Money Hysteria is entirely possible. In fact, this is a rather strong potential. If it appears, IT IS MOST LIKELY TO BE DOLLAR SCARES. Any progressive "weakening" of the dollar may collide with other forces, such as the possible switch of some international oil production from dollar prices to euro prices.

Since the value of the dollar is very much dependent upon the fact that you can directly buy oil with it, the weakening of the dollar's ability to buy oil directly will cast doubt on its long term power. In effect, it will seem like the dollar is no longer a stable reserve currency and this may trigger hysteria to dump the dollar in favor of euros.

This could easily become too much too soon and it could cause so much chaos that the stock markets might nose-dive sharply as people seek cover in euro-priced shelters. To prepare for this proviso, you might want to set up euro savings accounts. Consider saving your "cash" as euros in affiliates of European Banks.

Problem #5: Domestic Politics

It is probably wise to move to the Bear Scenario about the time it appears that Bush's poll margins are dropping steadily and three or four serious parties can be seen running in the 2004 elections.

Regardless of who wins the 2004 elections in the U.S., the fragile economy is not likely to improve enough to sustain a real boom in stock values. Corporate expansion strategies are primarily directed at international expansion and flight of job expansion capital out of the U.S. Only a major shift into an intensive National Economic Industrial Policy for job expansion in the U.S. will provide enough confidence to ignite a real economic boom and a rise in stock values.

It is probably wisest to assume that the 2004 elections are too soon for Americans to "get their act together". They will need to get coherent on a large scale about who they are as people so they can forge a new National Economic Industrial Policy. This will take some time and strong factions are likely to keep people divided until the hands of enough people are forced by economic circumstances.

Under this assumption, the main structural problems in the U.S. economy will not be addressed until after the collapse of the economy during 2006 and 2007. Accordingly, a DJI of 10,500 – 11,500 or so will begin a final precipitous decline in 2005 or 2006 and sometime in 2007 or 2008 will bankrupt out a great many American equity positions at $.10 on the dollar.

How It May Play Out

It is very hard to predict which one of these provisos is the most dangerous. All or none of them could come into serious play. As well, a small number of major players can profoundly alter the course of these provisos, or even the general flow of international affairs, as 911 quite clearly demonstrated.

Thus many factors could cause the timing on the collapse to accelerate, or it could cause it to be delayed somewhat. Accordingly, the projections and timing points in this Section must be considered to be highly hedged and qualified by unknowable factors.

For at least the next 12 months, a return to confidence and rising equity values will require that the invasion of Iraq be managed in a way which shows some ability to overcome political destabilization in the Arab world. This will not necessarily be easy. There are many snakes in the grass and some of them are American. There are prominent members of the Imperial Faction who WANT, even as Al Qaeda and Hezbollah want, the Arab world to destabilize as rapidly as possible, the better to rationalize even more American control over Middle Eastern nations. Some members of the Imperial Faction want to take over and transform the nations of Syria, Iran, Saudia Arabia, Libya, and the Sudan.

But it is doubtful that such ambitions can come into play during the next 24 months, the U.S. can barely occupy Iraq adequately. At the moment, the "power greed" of the Imperial Faction has bitten off as much as they can possibly chew and they will be busy for quite a spell. Accordingly, no major new war initiatives are likely prior to the elections of 2004.

If destabilizing forces in the Arab world are overcome and the economy becomes robust enough by 2004, demand for American equities may support a bull run surge in the market. It could foolishly boost the DJI past 10,000 during late 2004 or during early 2005. The moment it pushes 10,000, when every stock tipster, promoter, and hustler is telling the world to buy buy buy, this is probably exactly the moment to cash out entirely. At that time sell all stock and bond equities, assume that the country will be bankrupt by the end of 2006.

If this rosy projection expansion occurs, Bush may win re-election. He emotes a studied "sincerity" quite well. At least half the population is effectively absorbed within the Manifest Destiny Bubble under hypnotic control of the Mass Broadcast Media. They believe Bush's whopping Big Lies and they will vote for him to reaffirm their faith.

This may give Bush enough of a solid block of predators and sheeple that he out-polls an opposition which is fractured between Greens, bolting Liberals, and faux-centrist Democrats. If so, the Imperial Faction is likely to initiate an expansion of its war plans during 2005, probably aimed at one of the countries which borders Iraq. Under this condition, don't be tempted to remain in the market. Make sure you liquidate before more war talk gains enough headway to depress the stock markets.

The Crash Of 2006 And The "New Deal In 2007"

To recap the ***Modest Bull Bubble Scenario***, the most we are likely to get during the next two to three years is a relatively short, shallow bubble of growth, a lot of instability and political controversy, and, regardless of who wins or loses the political races, a growing lack of confidence.

To move beyond this, we probably need what is technically known as a National Economic Industrial Policy. We need a heavy gauge all-seasons type of the kind used by European governments, Japan, and many other nations during the past 60 years. Most nations, with the U.S. being the biggest exception, once used a coherent, well coordinated social development plan to revive domestic industries and nourish them one by one to maintain and expand employment. They got into this habit as a matter of dire necessity for recovering from major wars, beginning with World Wars I and II.

America's time for such a plan probably has come and indeed it is a no-brainer to predict that American politics will take up this question of a national economic development plan with greater and greater concern and with louder and louder conversations during the next five years. These conversations are already buzzing in rising crescendo among populist, progressive, green, and liberal circles.

Indeed, it may be time for all nations to revive in concert a new National Economic Industrial Policy for the 21st century to replace the various "globalism" programs which have wreaked such economic and ecological disaster in many areas of the world.

The conversation about a NEIP conversion will be thrust into the elections of 2004 with strong populist currents and radically new political agendas coming to the fore through a number of groups and new candidates for national positions. Both political parties are likely to experience some very choppy waters with major debates and strong disagreements between the various wings of the parties over the elements (industries) and scope of an NEIP.

The movements to throw out the Imperial Faction and realign American government will come from both the right and the left. It may even come to pass that a viable new militant populist coalition may be created drawing converts from both liberals and conservatives who are equally disconnected from the traditional parties. Probably not in 2004, but most certainly a populist/progressive coalition will begin to wax ascendent in 2006 and will take control of the American government under dire, virtually proto-revolutionary conditions during 2008.

The greatest fireworks will most likely occur around the Democratic party. Liberals, like the left in general, are generally highly incensed by the wholesale corruption and bald lies of the Bush Administration. They are emotionally highly motivated to mount a very aggressive campaign to dissolve away faux-centrist office holders and candidates in favor of strongly expressed positions about how to rescue the U.S. from Republicrat predators and the prevaricating incompetence of the Imperial Faction which has seriously compromised the international standing of the United States.

How the electoral politics will all play out no one can predict. Bush still holds enough cards to play that he could still win re-election in 2004, though if he does it will be in a razor thin squeeker like 2000. Or, a new President may be elected by a landslide. Too much hangs in the balance of the free will of individuals to call the outcomes of 2004 at this writing (July 2003).

If we are lucky, regardless of who wins or loses an election, a serious national debate will begin which deflates the Manifest Destiny Bubble, begins a re-trenchment from the reckless adventurism in the Middle East, and sets in motion a serious quest to define a national economic development policy.

Such a plan will not help us much during the next two years. Even if we tried tomorrow to create such a plan it is doubtful that such a plan could even be evolved in anything less than a couple of years. Let us hope then that we are lucky enough to have a brief revival in the economy during the next two years without a serious plan.

Even with this grace, we will barely have enough breathing room to prepare for the crash of 2006. The national conversation on this policy should begin yesterday. A highly progressive NEIP will be an absolute requirement for working our way out of the crash of 2006. Most likely, only a deep, thorough, systematic plan fully the equal of "The New Deal" will be adequate.

This need will be accelerated mightily during coming years by the simple fact that Europeans do not need the American economy. As the global economy of Europe and those who trade with it convert to the use of the euro, the dollar will lose more and more "use" and thus "value" as an international trading mechanism. This will be most apparent with the tendency of the euro to gain in value vis a vis the dollar, as it is already doing.

The only fix for this is a return to the creation of real industries in America which provide real jobs for Americans at traditional "journeyman" rates of prosperity. If the best we can do is create $7.00/hr part-time sans-health-plan jobs at Wallie World, we might as well hang it up and migrate to Mars. Life on Earth just ain't worth it.

A serious industrial policy to transform the economy for the long run requires that the Americans get real about who they are and what is really happening. This of course will need to be fashioned into an ideology to guide the use of government to accomplish ends which are frankly far more egalitarian in orientation than suits the taste of America's free-booting, jet-setting, corporate swindlers.

One of the major planks of such an industrial policy MUST BE to disconnect politicians and the conduct of governance from the corporatized syndicates. Wall street and national politicians talk free enterprise and the wonders of capitalism to secure votes. This talk goes over well in the Heartland with small business owners and self-employed professionals who think these things are meant for them. But the corporate syndicates, masquerading as a Newt Gingrich or a George Bush, or the FCC, or even as a Fox Cable Channel, consistently turn around and bury a million small businesses in a blizzard of increasing cumulative tax burdens, liabilities, regulations, and denial of opportunities.

In the hands of America's Century 21 sorcerers, "capitalism" and "free entreprise" have mainly been reduced to Orwellian terms which are used to euphemise the concentration of power and opportunity into a small network of crony syndicates. The "lords of the manors" run increasingly huge feudal-like estates in which the comport themselves as arrogantly as they please in a rising tide of one-sided contracts maintained by an unlimited slushfund tended by bagmen who buy and sell as many elections as possible.

One way around this is to provide strong remedies in a NEIP. One of these one remedies may be a coordinated national plan to institutionalize a strong return to small business roots by funding programs which foster and encourage small business and self-employed entrepreneurial activity.

This strong "main street" commitment would help overcome one of the key weaknesses in the New Deal Coalition. Early on, the Democrats focused on reforming mega-corporations and developing worker benefits within their frameworks. Democrats felt this would give them the biggest social bang for their efforts. It was a lot cheaper to oversee 100 corporations which employed 10 million people than it was to oversee 100,000 corporations which employed the same number.

That made sense some 50 years ago but the focus on corporations no longer pays much "social dividend". Though corporations receive the overwhelming bulk of government largesse, contracts, and markets, they create only a small percentage of new jobs. And for their efforts to provide a government which fostered the mega-corporations, the Democrats mainly now get unceasing right wing negative attack ads against Democratic candidates at every election, paid for out of the largess of the crony capitalists who have seized control of many of the major corporations.

Democrats need to return to the deepest and most productive roots. Small scale entrepeneurial activity, multiplied across the continent, now provides the largest job multiplier for every dollar spent. (See David Birch's work on entrepreneurial activity since the mid-1970s at MIT.)

Within the current political context, this kind of thinking is of course exceedingly remote. To insure that wealthy elites receive disproportionately huge shares of new wealth, a powerful minority of Americans have enshrined a Republicrat theocracy through most of the mass media, the electoral machinery, and American government. As of May 2003, the Federal government is of course nominally controlled by the Republicans. A National Economic Industrial Policy structured on egalitarian principles and "main street" roots is diametrically opposed to the prime imperative of this class.

Which is? Free enterprise for sociopaths who are clever enough to masquerade as crony capitalists. Somehow, by amazing media sorcery over the past 20 years, ordinary Americans and conventional investors have been duped into believing that corporate leaders deserve, in some mystical way, some astoundingly Gevaldt! levels of preposterous salaries for transferring production overseas, eliminating pension and health plans, and driving their domestic companies into bankruptcy to be recycled into another conglomerate pyramid scheme through which Wall Street sorcerers can puff up another soufflé of froth which they can dump on suckers for 1000 percent profit margins. From the actual results on the ground in reality, this pretty much appears to be America's current social development plan.

Even more amazing, Americans continue to buy the bill of goods that these pirates understand the economy and know how to fix it with the right policies. The truth, as Sherlock Holmes used to say, is elementary. The economy is broken and the corporated free-booters have not a clue how to fix it or they would. So instead, they are merely doing what they always do, they are picking the bones clean. Accordingly, given existing control of the U.S. government by the "Republicrats, any coherent industrial economic plan will be mugged in the halls of the Capital Dome and left to rot in the bushes out back.

In other words, the major economic reform and stimulus programs which are required to truly fix some of the gaping "structural" holes, efforts far beyond what is currently in process, are unlikely during the next two years. For that matter, they are unlikely during the next five years, though some serious starts in that direction may be possible after the elections of 2004.

Accordingly, after some modest piddling expansion this year and the next couple, a huge depression wave may begin to flow through the economy, perhaps from as early as 2005, certainly no later than 2006, and continue on through to 2008, just like Cayce's Depression Cycle predicts.

If so, here is the good news! American equities, values, and effective wage rates may fall to as little as $.10 on the dollar compared to their current value vis a vis the euro and a barrel of oil. This is not necessarily bad. It's absolutely great if you don't have much stake in it. The ultimate Bear position is that this is very very good. This contraction will bankrupt a vast portion of today's self-serving elites and remove their abilities to manipulate the mass media and the voters.

Lies by the exploiting political factions to preserve the status quo for the latest corporate scams won't work nearly as well when people are mired in their sudden descent into poverty. People have an innate ability at such times to tune out everything which is not working for them.

Progressive social movements will suddenly gain more power to institute a "New Deal" than they have possessed since 1936.

Focusing The New Center In The Spiritual Reality

It is sad to say, but the American public, collectively, like all publics, gives the distinct impression of being about as stupid as a mule. Accordingly, it probably needs the Missouri 2x4 applied strongly against its hindquarters to induce movement in the desired direction. God provides this every once in a great while in the form of major depressions and recessions.

As a steep depression sets in, a massive destruction of paper values spreads like a bitter medicine, waking people up from a generation of illusions and dillusions. Suddenly they become more susceptible to reason and issues of equity, less prone to the self-serving claims of yesteryear. This bitter medicine is probably essential before a new socially-regarding majority can emerge to elect people of genuine merit who really care about creating social justice.

That is pretty much how it worked for the FDR coalition. And it worked very well. Essentially it was idealism mixed with pragmatism in those 1930's and 1940's "little people" which engendered the basis for this past 60 years of American economic, social, and cultural growth. It was NOT the agenda NOR the efforts of self-serving corporate elites and their fancy-pants academic rationalizers which saved the day.

Some people think that the elections of 2004 will offer some fireworks and the beginning of some positive changes. This may be so, but this will be merely the prelude to the elections of 2008. There probably will be too many fouling dregs in the system in 2004 to "break clean" in the manner which is necessary. Thus it may take until 2008 to break free of the corrupting and repressive forces of the past. At that time, it is quite possible that three fourths of career politicians will find themselves replaced. So it was during the 1932 and 1934 elections.

Over this issue of industrial policy, then, we will see the growing polarization arise. Indeed, it is already arising. On the one hand we have the churlish, secular forces of extreme individualism fostered by extreme competition, TV violence, and games of dominance/submission. On the other hand, we have the social-regarding or spiritual forces which are centered on "God Game", including a large diversity of people expressing principles of higher meaning, morality, ethics, and the real manifestation of spiritual love and mutual concern for all people.

These spiritual themes could be seen proudly and profusely displayed by the peace marchers in February and March of 2003.

Though the groups are diverse, ranging all around the globe and speaking dozens of different languages, and even though they belonged to what some people have supposed in the past to be divergent philosophies, they all easily united on the simple plank of a common concern for each member of humanity and a refusal to consider subjecting women and children to bombs and machine gun fire from rolling tanks.

On this point, on this mutuality of respect and concern, on this spiritual point, all of the politics of the next few years will increasingly turn. Increasingly, all humans will be forced by catastrophic collapse of their economics to come to terms with how they will take care for one another and insure a world which works for all within a real framework of ethics and morality which respects the great religious quests of the past.

Moral and philosophical relativism within a purely secular framework of consciousness cannot possibly find the way. For one thing, it is not nearly imaginative enough, but this discussion is better left to another day. Those on the left who are uncomfortable with this spiritual point will engender more confusion in their activities than clarity.

Mutual concern within a spiritualizing framework, making the world work for everyone, the task begins here with this focus. It is this simple and because it is this simple, it is real. This is the only reality. All of the economic theories, philosophies, praxis, political parties, jihads, bigoted bully boys, and legions of experts be damned to perpetual frustration till they come to this elementary point which connects us to spirit, to love, to higher purpose, and hence, by any name, to that which people call God.

Mutual concern, the spiritual connection, making the world work for everyone. On this simple point, the dialogue for World Civilization will begin. The debates have already begun and the first battles will be fought in 2004. Great political controversy will ensue through to 2008. Then, over the corpse of the American Republic, a great political war will be fought between coalitions of the same forces which struggled in ‘32 and ‘36. Best have some safe ground and nothing in paper.

Section Six

The Coming Dollar Panics

Many nations in Europe and other quarters of the world are beginning to experiment with how to overcome the past few decades of dominance by American cultural and economic institutions. This shift in consciousness and alignment of politics is now accelerating at a brisk clip, directly fueled by the polarizing ways of America's Imperial Faction.

As the world realigns into a multi-polar world, a vast portion of Eurasia is tuning into the euro as a potential preferred world currency standard of value among many nations. Already, some non-European nations are eliminating the dollar as a reserve and trading currency in favor of the Euro.

As this attunement accelerates, American politics and economics could be severely afflicted by increasing economic imbalances and weaknesses. The biggest source of problems is likely to be the value of the dollar. Because of the severe structural problems and "***giant sucking sounds***" in the U.S. economy, the stability of the dollar is losing confidence and it is becoming less and less desired for trade by many international forces.

Indeed, as of July 2003, some non-European countries have already switched from the dollar to the euro for conducting their international banking. This has prompted some financial circles in Europe and the U.S. to openly promote the idea that the euro is going to displace the dollar as a reserve currency in many parts of the world while the dollar rapidly declines in value, perhaps by as much as 25 to 40 percent below its June 2003 value. To listen to some brokers in the U.S., the panic has already begun and it is imperative for investors to buy euros immediately.

Obviously the desire of brokers is to sell. For those who happen to be selling currency, the euro these days is an easy sell if they can convince their clients that the dollar is doomed to a come-down. Accordingly, as a few phone conversations with brokers easily reveals, many circles are busy selling the notion that international confidence in the dollar is fading away. Dump the dollar, they loudly suggest, and save your wealth in euros.

That of course would do the trick, if everyone did it. A dollar dumping mob could ruin the dollar in very little time. But how likely is this? After all, the dollar has been highly stable for 30 years with only slow changes in its average value. The whole world trades in the dollar and most major central banks of foreign nations use the dollar as a "reserve asset" to establish their credit with the IMF. The last thing most international traders want to see is a decline in the value of the dollar. It almost automatically insures a lower value for their own currency. Surely, won't these institutions work to hold the dollar on an even keel?

They probably will. Yet, what if that is not sufficient as some Bear analysts now claim? What if confidence in the dollar really is being lost, and and what if a major panic over the dollar sets in? What does this panic look like? How might it occur? Won't this traumatize the American economy and greatly disrupt its people? What will this do to the struggle to fix the structural issues? How can we protect ourselves both individually and collectively?

To answer these questions we need to begin once again by looking at the historical context to reflect on the long term trends.

The Eclipse of American Fixations

We need especially to perceive that the most important long term trend is the displacement of the "American point of view". This "American point of view" of world events is more or less the "slant" and the "selection" which is proffered up by the Boomers in the Mass Broadcast Media and the major metropolitan newspaper headlines. This "world view" reigned for some 50 years during the cold war and Clinton years, probably most of the time with far more influence than it deserved.

Though largely dominant in at least half the world, the "American point of view" is now being eclipsed more and more in many nations. As America's agendas and fixations command less and less attention in nations around the world, the U.S. government is finding it increasingly difficult to obtain attunement with its goals and activities. Trying to sell cooperation with unilateral made-in-the-USA moves is becoming harder and harder to do.

This displacement process has been slowly accelerating since the collapse of the Soviet Union. The process was rarely directly visible among international political circles but a progression of unilateral U.S. moves and various ideological statements by the Imperial Faction about the preeminant role of the U.S. has made the displacement increasingly visible since the Fall of 2002. By April 2003, just barely a year since the beginning of the U.S. war drive into Iraq, the displacement was highly visible in the policies and actions of major European nations, as well as in the public opinion of most nations of Europe.

Does it not seem obvious, then, during 2002 and the early part of 2003, that the international "news fixation" on the American point of view has broken down? Global points of view, with many attitudes at variance from the traditional American slant, are beginning to become more noticeable in all things. From these trans-national points of view, a growing world re-alignment is rapidly forming a vision about the United States which is at substantial variance from the image which American Media has broadcast worldwide for the last 50 years in a variety of formats, from radio and TV to film and books. The realities of American imperialism, for instance, are beoming more widely discussed than the illusory fantasies of the "American Dream".

This shift in consciousness is observed most strongly in Eurasia. It is now clear that the Eurasian powers are forging closer relationships among themselves than ever before. The French and Germans have rekindled a sense of international destiny around strong policies which support humanitarian values and which aim for peaceful solutions. Together with the Russians and many other European nations, the Euro peoples are discovering all the reasons why a strong Eurasian alignment (and value of the euro vis a vis the dollar) really is the best natural order for their affairs. Into this growing "euro" orbit, the Russians have been effective in pulling China into parallel alignment.

In a perverse way, this growing euro orbit will be greatly accelerated by the on-going activity of the Imperial Faction. By making the United Nations "irrelevant", at least temporarily, the Imperial Faction is effectively making the Euro Union a mandatory counterweight to the unbridled power of the United States. Even as "Manifest Destiny" Bush has proclaimed his doctrine of preemption, Russia's Putin has in various discrete ways let be known the need for the Eurasians to hang together lest King George and the Empire hang them separately. Accordingly, in more and more ways, these peoples and their governments will band together more and more.

From these peoples and their institutions will come a range of challenges to the activities of the United States. Before long, even Blair and the British, who really do want to believe in the United Nations, are likely to find Europe more congenial than the overweening ambitions of the Imperial Faction. Some of these challenges will directly address the issue of American violations of the United Nations Charter and various provisions of international law. There are a great many violations. Some of the challenges will directly address, in ways which increasingly confront the U.S. economically, the refusal of the United States government to deal with a range of important environmental, social, medical, military, and economic issues in accordance with international standards and treaties.

Not far behind in the challenge sweepstakes will be acute international political activity related to the environmental destruction and progressive poisoning of all living beings which is being created by many consumer and military product streams. Not the least of these issues is the "depleted uranium" metal which is used to construct America's bombs and shells. And close on now is a rapidly rising awareness of the magnitude of unwanted changes and very strange side-effects which American bio-engineering and pharmaceutical companies are creating by championing their "Franken" products and their little shops of horrors in their drug patent monopoly divisions.

For a variety of reasons, one of which is simply the greater density of population in Europe, a well developed labor intensive agriculture, and traditions of higher quality of taste, Europeans are widely contemptuous of American environmental policies, use of chemicals, and attempts to bio-engineer radical changes in plants and medicines in the quest for patent monopolies.

This expanding European awareness is also awakening to realities about the injustices in the world created by international trading and banking practices. Championed by U.S. corporate and moneyed interests, many international economic institutions are predominately managed out of the U.S. by elite American circles. Some of these international practices and institutions are ruining economies in the world which, like the U.S. economy, were once prosperous. After 50 years of IMF and World Bank activities throughout the Third World, poverty and violence in the Third World is increasing, not decreasing.

During the last 20 years, as in America, the so-called "middle class" has shrunk and the after-tax earnings of labor has contracted significantly in countries which are being "stabilized" by the International Monetary Fund. Brazil, Argentina, and others have already hit this wall. Japan appears to be barely avoiding the wall but the U.S. may be approaching so close at such a rapid clip from behind that it may shove Japan into the wall when it too finally hits the wall.

Political movements around the world, especially in Eurasia, are increasingly active in loudly demanding a radical re-alignment of these international institutions. Given the increasing rate of economic failure in the world, none of these movements are going away any time soon, they are likely to grow ever more influential in causing international meetings to be held in out of the way places.

They are likely to be increasingly successful in forcing changes in lending and monetary policies, especially through a diversion of international lending from private debt banking to new forms of socialized expansion banking.

American isolation from Eurasian alignments of world opinion will continue to grow so long as policies remain in the U.S. which support the social and political dominance of free-booting corporate-based swindlers who masquerade as the CEO's of "public companies". A great many of those who came into prominent notice with the crash of the high tech and dot-com stock bubble in 2000/01 have personal ethics which stink which such a fetid smell to most of the world that it is a wonder that Americans have not declared martial law and seized some of the public corporations to purge them of corruption by these country-club mafioso.

With this garbage rattling loudly in their closets, and with legions of corporate rats abandoning ship by transferring equipment and jobs overseas, Americans increasingly will find they have a weak hand in dealing with Eurasians. They will find that their hand will continue to deteriorate with each "draw" at the negotiation tables. They will find that they are commanding less and less "confidence" and that their positions are more and more isolated.

Since economic recovery will require international cooperation, this loss of confidence will become more and more of a problem so long as Americans are seen promoting merely self-serving policies which favor narrow elite interests in the U.S.

Dollars, Euros, & Oil

The eclipse of the "American point of view" in the world will be accompanied by the eclipse of the pre-eminence of the dollar. As the euro markets expand and bankers begin to develop their asset bases in euros, the dollar will fall more and more into a long term downward slump.

Some officials in the Imperial Faction have already experimented with the idea of consciously forcing the dollar down in value to stimulate the American economy into new growth. Quoting an unnamed "high ranking diplomatic official" in the Bush Administration who had attended a meeting in Brussels with European governments,

The New York Times rumored the end of the old era of the high-priced dollar and the coming of the “cheap dollar”.

This new policy, if true, would reverse nearly a decade of economic policy to allow the dollar to slide in value and make it cheap vis a vis the new euro. The objective was to increase exports, reduce imports, reduce the U.S. balance of payments problems, and stimulate growth in the U.S. economy.

The story made sense, at least from an American point of view. Since interest rates have been shoved to as low as they can get, obviously this old tried and true domestic arm of monetary policy is not working. Central bankers are practically giving the money away to banks but banks can't find anybody new to lend to who isn't bankrupt or without income.

Seemingly it is now time to turn to the international arm of monetary policy. It is highly logical now to cheapen the dollar, which allows foreigners to convert their currency into greater numbers of dollars. Hopefully this will permit foreigners to buy and invest more in the U.S., which of course will stimulate the U.S. economy to expand. Think of it as a form of international “sale” to boost turn-over in the U.S. economy.

The story may have made too much sense and thus may have caused too much alarm in Europe and in the U.S. Rumor or not, the story worked and during most of May the dollar slid downwards in value and lost about four cents against the euro. To halt further erosion, the U.S. government began to back pedal during June 2003. Bush Junior announced during his G-8 Summit Meeting in Europe on the first of June 2003 that he was committed to a “strong dollar” and would act appropriately.

This would seem to over-rule the whispers which came from other mouths in the Imperial Faction and which were turned into megaphones on the pages of the New York Times. Or maybe not. Bush Junior and his minions are by now infamous for double-dealing out of both sides of their mouths. Bush may have decided to “appear” for a while to be saving the current relative values of the currencies, in the same fashion that he decided to appease the Europeans by “appearing” to go through the United Nations to finish the disarmament of Iraq.

Regardless, he did a good job of “appearing”. News articles about his commitment to a “strong” dollar appeared in all the wire services and major newsboards on the internet. Since then, in the course of four weeks, the dollar halted its slide and rose about three cents vis a vis the euro.

Europe's commercial interests were most likely strongly aghast at seeing the euro gain in value over the dollar and thus most likely so were Europe's primary politicians. To save some face, indirectly, for the amazingly frosty ways he has dealt with Europe up to this point, Bush may be waffling on the Dollar Policy and he may continue to do so for a period of time, perhaps for so long as he needs to gain European approval of the latest version of the Palestinian Peace Plan the U.S. government is sponsoring.

Keep in mind that Bush Junior made a point of observing that the U.S. government has an "independent" central bank (the Federal Reserve). Was this his way of providing a personal "out" if the Federal Reserve "independently" acts in ways which are permissive of a cheap dollar? Is this part of a waffling dodge 'em routine?

Such questions, which are highly legitimate considering the habits of this Tall-Tale Telling Texan, will keep the currency poker game very interesting. Will the Feds support an expensive dollar, or some decline in value over the course of the next two years? Only the Shadows lurking in the halls of the U.S. Treasury know.

The "Cheap Dollar" policy, if indeed it is a policy, has already worked quite well. Two weeks after the New York Times article appeared about the middle of April 2003, the euro increased in value by 2%. That is enough to sit up and take notice the next time you hear such a story in such a source, especially if you have U.S. dollars in bank accounts which are not earning much more than that for an entire year.

It would seem that if you want a good investment for the year or years ahead, buying euros and holding them for a couple of years will probably be about the safest, possible way to make money with your money. Some brokerage houses and investment gurus are already touting the possibility of making 10, 20, even 40 percent with your money in euros, perhaps even more.

It should not be much of a leap to suppose that a lot of Americans will get the same idea and buy up more and more euros, no doubt adding to a constantly rising value during the next couple of years, perhaps for longer. Since money stashed in American banks is a dead letter, what better way to give it life than to buy euros and stash it in a European bank which pays a higher rate of interest even while the value is appreciating at a rapid rate.

Mom and Pop may not flock to the idea, but corporations with cash reserves of a few millions of dollars might very well do so. There are such huge quantities of dollars floating around internationally, it should not take much effort to stimulate people to convert a lot more dollars into euros.

From an economic stimulus point of view, this might seem to be an effective idea which is easy to implement. But, because of the role of the U.S. dollar as an international standard of value, it is not so easy as it might seem to adjust its value relative to other currencies.

A cheaper dollar may produce positive effects during 2003 and 2004. But this sort of unilateral policy can easily produce too many uncontrollable effects which provoke a deepening crisis. Indeed, this program may be the final watershed for the chain of causation which will bring the collapse of 2006, quite the opposite of what the policy is intended to produce.

A downward surge will create unintended consequences far and wide. These could easily spiral out of control and create more harm than good. It could even create a dollar panic, or two or three, a sudden demand to convert dollars into euros which thus suddenly causes the value of the dollar to drop radically.

Think of the effects like the "broom straws" which the Sorcerer's Apprentice (from the Disney 1930's classic animation) inadvertently created. The effects may independently spiral out of all control to flood the U.S. with a severely devalued dollar, "paper equities" which no one wants, large numbers of corporate bankruptcies, and even tens of millions of jobless people banging pots and pans in the streets of America.

Also like the straws of the Sorcerer's Apprentice, once this is set in motion there may be no way to retract the policy or gain control of it through further manipulations. Once confidence in the game is broken, the game is up, there is no return.

To watch for this and prepare for dealing with it, we need to learn how to observe the forces and spiral movements which are in play around the dollar. Most importantly we need to tune into the international political factors which may cause a precipitous drop in the value of the dollar.

This danger is now substantially greater than at any time in history. There is a rapidly growing possibility that oil producing countries will begin to switch the pricing of their oil from dollars to euros. This of course will reduce the dollar's use, which will lessen its value while increasing the value of the euro. If the oil producers all switched at once that would be quite a sudden and large drop in value of the dollar.

OPEC won't do it precipitously because they fully understand that it is against their self interest. One of the great strengths of the dollar at this point in history is that almost all oil is denominated for sale in dollars at a price which does not change much with fluctuations in the values of other currencies. This in effect means that the dollar is "backed" by oil. Accordingly, any company, bank, or country can accept dollars in payments because they know exactly how much the dollar is worth vis a vis oil.

This convenient utility of the dollar is the reason why it is known as the "petrodollar". Since the petrodollar can always buy oil anywhere, it is safe to stockpile huge quantities of dollars and use them as "reserves" to capitalize both central banks and private banks. By this convenience, gold and silver as the backing for money disappeared a long time ago. To preserve the dollar's relative value in the world, it is more than enough that dollars can be converted into a barrel of oil.

Naturally, all the banks, including the major oil producing countries which sell in dollars, want both stable oil prices and a stable value for the dollar. Life is good under those circumstances and so the world has a vested interest in maintaining the value of the dollar. But everything begins to get strange and even sour when oil prices and the value of the dollar begin to shift around.

Accordingly, one of the main functions of central bankers has been to make adjustments to keep the dollar at a high relative value vis a vis other currencies, including their own. This allows them to hold onto the value of their "dollar reserves", and this maintains the ability of their countries to import the same amount of oil with the same amout of export earnings. This of course tends strongly to keep their own economies on a stable basis.

If the value of the dollar falls lower, they can't buy as much oil, and they have less "reserve" value to back their own currency. They then can't loan as much, nor borrow as much, and their own economy is threatened with unwelcome contraction and recession.

Think about the dollar as an ounce of gold. When the price of gold sags down, it reduces the value of their inventory. So they go buy more to increase the value of their inventory. And this of course raises the value of gold because the "bargain priced" gold quickly disappears. Same thing with the dollar.

There is more incentive to buy cheap dollars up, thus making them disappear, because they are a lot easy to deal with than a vault of gold. With electronic transfers of credits, those dollars can be sent anywhere in the world to buy oil anywhere in the world. Dollars equals oil, no questions asked. That's why the international monetary system works with dollars and not with gold. Gold is totally unnecessary, completely atavistic.

Slow changes bit by bit over a long period of time in the prices and values can be easily accommodated and in fact have been. But if the rate of change speeds up, eventually countries will have a strong incentive to convert their unstable, shrinking dollar into something else which is easier to deal with. If they begin to do so, the value of the dollar will suddenly sink, sink, and sink some more, quite possibly igniting a world dollar panic. Rather than everyone having a vested interest in retaining the stability of their dollars, everyone will have a terrible disincentive to keep them. Panic could result just like the old run-on-the-bank panics. But instead of there being a run on American banks, there will be a run on other banks to buy other currencies.

At a practical level, this has been more or less impossible until now, nothing has been available as a substitute for the dollar. The Japanese began talking about a Yen world during the late 1980's to replace what some Japanese saw as the inferior dollar, but since then their banking system has essentially collapsed under the weight of all the bad loans they made. The current value of the Yen is quite fragile, to say the least, which is more or less a charade by which the Japanese establishment pretends it is not bankrupt. This of course allows a small elite of bankrupt idiots to continue to pretend that they are more deserving of power and wealth than their countrymen.

But now, the euro is available. It has the backing of a half billion people and by far the largest economic block of resources, manufacturing, labor, and educated people in the world. Combined, the powerhouse of Europe with Russia exceeds the powerhouse of the United States. In contrast to the U.S., the nations and the people who use the euro actually manufacture great quantities of goods for all levels of all markets.

With Russia in cooperation, the euro block has all the resources needed for anything, except, they need to import oil, metals, and tropical food stuffs, fibers, and spices.

A country like Iran, or any other major oil producer for that matter, could easily find at any moment that it is well worth their while to "sweetheart" their oil to euro values so that they can more easily build up their bank "reserves" in a currency, the euro, which is appreciating in value, rather than keep dollars, which are losing in value. Such a move would immediately weaken the utility of the dollar and begin a general movement of other oil producers and dollar reserve holders to do the same thing.

It's coming soon. Iran has already converted its foreign reserves to euros, which means it no longer keeps dollars nor pays for its imported goods with dollars.

It could spiral out from there. Such is how the panic would begin. As of May 2003, the euro has already appreciated some 30%

since its introduction two years ago. On the strength of newspaper rumors, it escalated some 2% in less than 30 days during April 2003, and at the rate of about 1% each week during May 2003. Since then it has gone down a few pennies at the express promise of Bush Junior to maintain the value of the dollar.

What can we gain from this? There is a lot of tug of war going on in the world over the petrodollar and it is only as stable as the main decision makers decide to make it. None of them has complete control, and the current balances could easily change, rapidly, on the basis of one single newspaper article. Be diligent. Be very diligent.

No doubt the pressure and the push towards a cheaper dollar has been accelerated by the precipitous loss of American credibility which was engendered by the invasion of Iraq. By invading unilaterally before United Nations operations on the ground in Iraq could be completed, the Bush Administration broke a variety of treaty obligations and international laws and thus rendered its "reliability" and "statements" something dramatically less than "credible". Why hold dollars from a government which is blatantly dissembling, amateurishly manipulative, and is turning once friendly nations into distant ones?

It is will be interesting to watch during 2003 and early 2004 how well the U.S. government is able to maintain its "strong dollar" policy in the face of so much international polarization. With unilateralism and a history of other one-sided treaty renunciations, the Bush Administration has nearly perfectly set the stage for failure with hardened polarized attitudes in many nations and a lot of passive non-cooperation.

Unfortunately, as is abundantly clear everywhere one turns, the Bush Administration is managed by people with very little long range historical and cultural perspective. It is highly doubtful that they have the cultural sensitivities and historical acumen to undertake a reliable dollar policy. It could come to pass that the dollar will blow up in their faces, as did the consensus of the United Nations Security Council. If it does blow up, it is likely to be in the form of a dollar panic in which the value of the dollar plunges against both the euro and a barrel of oil.

The Reserve Dollar

To see how the Dollar Panics are likely to play out and the precursor signals which will probably announce various bouts of them, we need to understand the historical background of the Petrodollar and the international monetary system.

As a result of the Great World Depression and World War II during the 1930's and 1940's, the U.S. dollar became the de facto world standard for the value of most currencies. It held that status partly because the central bankers of the largest industrial companies agreed to agree with each other that the U.S. dollar was worth the $35/ounce of gold that Roosevelt's bankers had said it was worth.

This had little meaning for most of the world as most of the central bankers agreed to only exchange gold and U.S. dollars between themselves, never to buy and sell bullion gold on open markets in a way which would undercut the value of the dollar. In effect, the bankers preserved an elaborate shell game, pretending that gold was under the dollar.

But it was not really. No one else in the U.S. could buy and sell gold for monetary purposes, nor could they convert the dollar back into gold, at least not in the United States. In effect, in one of history's benign scams, Roosevelt had taken the dollar off the "free market" gold standard and allowed its use only as a "marker" among the central banks of the major nations.

Since the central banks had a vested interest in preserving the relative values of their currencies, and since there was an enormous quantity of gold held by the U.S. Treasury after World War II, the world's central banks more or less froze the values of their currencies against the dollar and began to use the dollar itself as a marker in place of gold. If they needed gold, they figured they could always cash the dollars in to buy back gold at $35 per ounce directly from the U.S. Treasury. Since no one else except the central banks could buy the gold, it was a pretty safe bet. A small committee of central bankers could do whatever they needed to do over the phone or telegraph to preserve a tight control over values.

This worked very well for a generation and the system got formalized as the International Monetary Fund and the World Bank. But the system began to break down during the Vietnam War. The U.S. lost substantial political credibility in expending so much effort to eventually lose a brutal war, the Russians were producing and selling a lot of gold creating other streams of currencies and reserves in the world which could compete with the dollar, many more nations were polarized against the U.S. as a result of the Cold War, and there were now many more wealthy industrial players in the world all seeking to maximize their economic growth by exporting to the U.S. for dollars while buying as little as possible from the U.S.

So by the early 1970's, the system came under severe pressures. Nixon dumped some of the pressure by dropping the gold conversion value. Henceforth, gold and the U.S. dollar float freely and independently in value. This allowed the U.S. Treasury and the

International Monetary Fund far more latitude in adjusting exchange rates. Under this system, the U.S. dollar in effect became a "de facto" world reserve note.

Now this seems in the abstract to be perfectly worthless mumbo-jumbo, just plain voodoo. And in the abstract it probably is. This could not have worked unless the dollar had compelling uses, real power in the world. This it had in three forms. The U.S. had been and still to this day is one of the largest arms manufacturers in the world. U.S. dollars could be used to purchase large quantities of high quality weapons and munitions. The U.S. was and still is one of the world's most efficient growers of food. Dollars can always purchase huge quantities of surplus grains, quite often aggressively discounted under various types of so-called foreign aid programs.

Probably more importantly in the world of 2003, the greatest use of the dollar has come to be the ability to buy oil directly with dollars, no questions asked, anywhere in the world. To look at the power of this in greater detail, let's look through the logical eyes of Dr. Gavin R. Putland (Queensland University of Technology, Australia) who described the situation so lucidly in March 2003:

> "In 1973, the Organization of Petroleum Exporting Countries (OPEC) quadrupled the price of oil but continued to accept only U.S. Dollars in payment, so that the demand for dollars soared. From then on, the dollar was effectively backed by oil instead of gold -- and the U.S. government didn't even have to own the oil! Because dollars can buy OPEC oil, countries that need to import oil - i.e. most developed countries -- will accept dollars as payment for their exports. Hence everyone who needs to buy from those exporters will accept dollars as payment for other things, and so on, so that the dollar is the preferred global currency. To pay their bills, importers must have reserves of dollars. To prop up their currencies against speculative attacks, the central banks of all countries must have reserves of dollars. To get capital, poor countries must borrow dollars, and to service these debts they must export goods to obtain more dollars. About 2/3 of all currency reserves, more than 4/5 of all Currency transactions, more than half of the world's exports, and all loans from the International Monetary Fund (IMF) are denominated in dollars. As these things create demand for the dollar and shore up its

> value, OPEC is the more willing to accept payment in dollars. So the system is self-reinforcing."
>
> copyright March 26, 2003 Dr Gavin R. Putland,
> Queensland University of Technology, Australia,
> http://www.fair.org/press-releases/venezuela-editorials.html
> http://www.observer.co.uk/international/story/0,6903,688071,00.html

The catch is, the system is self-reinforcing only so long as the value of the dollar is relatively stable with other major currencies. But let's continue with Dr Putland's analysis:

> "The result is that America can export dollars, which cost nothing to produce, and receive real goods and services in return. As long as those dollars are spent outside America, they don't cause domestic inflation. And when they eventually find their way into foreign reserves, they can only be invested in American assets. This continuous flow of foreign investment (on the "capital account") props up the American real-estate market and stock market, and allows America to run a mammoth trade deficit (on the ``current account") without devaluing the dollar. America's imports now exceed its exports by almost 50% (or 5% of GDP) and its foreign debt is 60% of annual GDP."
>
> copyright March 26, 2003 Dr Gavin R. Putland,
> Queensland University of Technology, Australia,
> http://www.fair.org/press-releases/venezuela-editorials.html
> http://www.observer.co.uk/international/story/0,6903,688071,00.html

Slamming The Dollar

Quite obviously the entire logic would slam into reverse gear if oil producers collectively abandoned the dollar in favor of the euro. Both corporations and central banks would have to sell dollars to buy euros, causing the value of the dollar to sink. The dollar would buy less and less. To buy the same amount of oil as it currently does, America would have to increase its exports of real goods and real services to earn more dollars to buy more expensive euros. But American exports are decreasing (relative to imports), thus banks and corporations would have to dip into their save assets to make up the difference. Since a great many institutional "dollar assets" are held in the form of real estate and stock equities, institutions would need to sell more and more of these assets to convert to them into euros. This would create a buyer's market which could spiral out of control, plummet, and eventually even "crash" through the floorboards.

Naturally, all those who found themselves "shorted" by the deflation of values would tend into bankruptcy. In other words, any corporation with high levels of debt mortgaged on real estate which suddenly sinks in value, or on products which can no longer be produced because of high import costs, would find themselves defaulting. Jobs would vaporize, real estate markets depreciate, and banks would find their bad debts overwhelming their assets. The smell of massive bankruptcy would waft through the streets of even the most remote small towns in North America.

Dollars would disappear in large quantities from the U.S. but would circulate around the world in huge quantities, producing a dollar inflation. America would have to pay the entire cost for imported goods "on the dock" in real money, like as in euros, and that could be at equivalent dollar values which are several times higher than today's prices. Wal-Mart would find itself emptier and emptier.

Most likely the new world reserve currency of choice would be the euro, which, though only about three years old, is already taking hold in the world as a potent, highly regarded currency. Eleven members of the European Union (EU) adopted the euro as a common currency during 1999, Greece joined the euro a year later. On January 1, 2002, the twelve euro nations withdrew their national money from circulation and replaced it with the euro to make the biggest monetary reform in European history. Since inauguration of the euro, its value has appreciated by about 30% over the dollar, it began at about US$.85 and could purchase US$1.17 as of June 1, 2003.

The euro nations collectively have the powerhouse in their economy to support the euro even if it expands to serve as a major reserve currency. Euro banks offer higher interest rates than American banks and Europe as a whole does not have a huge foreign debt nor, MOST IMPORTANTLY, does it have a trade deficit. Both of these are two gigantic wounds in the side of the U.S. (This is where the giant sucking sound can be heard).

Even more strategically important, the euro nations conduct a bigger share of world trade than the U.S. and, because of their close physical proximity, they happen to be THE main trading partner of the Middle Eastern nations. Accordingly, the euro can inspire an easy confidence and requires far less external cooperation to maintain its relative value. Regardless of the dollar, the euro will fare well.

Given these circumstances, it probably will take little coaxing for Middle Eastern nations to switch to a euro pricing scheme for their oil. It is, regardless of what the U.S. does for good or ill, probably an historical inevitability that the euro will become the world's second reserve currency and eventually the main currency used throughout Eurasia. As the dollar plummets, the euro will fare very well indeed.

Possibly for these reasons and perhaps others, China started converting some of its currency reserves from dollars to euros during 2002. As well, North Korea abandoned the dollar and started using euros for trade. In December 2002, ten new countries were approved to become additional members of the European Union. They will begin to use the euro in May 2004. With these new members, the European Union will compose a common market of 450 million people, with Russia in close alliance. This market purchases over half of Middle Eastern oil.

This strong juggernaut will put strong pressure on Denmark, Sweden and the U.K., who are already members of the European Union, to convert their currency to the euro. It also gives strong logic to all OPEC nations to convert their oil prices and currency reserves to euros during 2004.

May and June 2004

This is what to watch for. Any OPEC nation which is thinking of converting to the euro will most likely begin like a cagey fox. If OPEC nations slowly convert their bank reserves into euros, following Iran's example, there will be little direct psychological impact on the "confidence" in the dollar. Shocks and upsets can be avoided.

Later, when they begin to price their oil in euros, doubtless a growing increase in demand for the euro will be immediately felt. This will cause a significant increase in the value of their euro reserves relative to the dollar in a fairly short period of time. No problem for banks with euros in the vaults. Big problem for banks with dollars.

This approach is most likely to be played in concert by States which are the most polarized from the U.S. Two OPEC members, the two most polarized from the U.S., have already dumped the dollar. It does not take an expert to guess which ones. Iran began talking about switching its oil prices to euros in 1999, even before the euro had tested its wings. Bush named Iran in his ``axis of evil" speech in January 2002 and Iran promptly converted most of its currency reserves to euros during the remaining months of 2002. In October 2000, Saddam announced that Iraqi oil would be sold in euros instead of dollars. Within a few months, Saddam converted Iraq's entire $10 billion ``oil for food" reserve fund from dollars to euros. These facts went unreported in the U.S. media.

The third prominent member of OPEC who threatens the dollar is Venezuela, the 4th largest oil producer. Hugo Chavez, Venezuela's twice elected president, hosted a conference in 2000 on the future of fossil fuels and renewable energy. The conference recommended that OPEC set up an electronic bartering system, to allow oil producers to directly trade oil for goods and services.

Venezuela set up the barter system to enable third world countries (those without large reserves of dollars or euros) to directly swap their goods and services for oil. Cuba took advantage of this offer and now provides free health services in Venezuelan villages in exchange for oil. Chavez arranged 13 other international swaps.

In April 2002, there was a coup against Chavez. This coup against a democratically elected President was welcomed by the Bush administration and by editorials in numerous American newspapers. But the coup collapsed two days later, leaving the wide perception that the U.S. administration had meddled inside the government of Venezuela and had encouraged the coup against the democratic will of the people of Venezuela. For its efforts to destabilize Chavez, the Bush adminstration has just been rewarded by Venezuela's switch to euro banking reserves and an offer to sell its oil in euros.

A great many people are still wondering just why the Bush administration had turned its back on traditional American democratic values and supported a gross violation of Venezuela's Constitution. Some say it is a jingoistic reaction to Venezuela's deal with communist Cuba. Others whisper that it is really about the petrodollar, in particular, about preserving its hegemony. Chavez threatens to undermine it. Many of the same voices whisper that the invasion of Iraq was also about the hegemony of the petrodollar, Saddam clearly did undermine it.

At all costs, many voices in the third world whisper, the Imperial Faction must preserve the dominance of the petrodollar to maintain the hegemony of the Century 21 American Empire, along with the widely hated International Monetary Fund and the World Bank which dictates banking and currency policies for most of the world even while administering a crushing load of unwise debt over most of the countries of the world which do not have their own oil.

So, about international economics, some say oil is the essence, the petrodollar the key, all else are merely excuses and convenient rationalizations. If this is so, the increasing adoption of the euro for oil pricing may provoke highly animated reactions from the U.S. government. What it can do to halt a consequent slide towards dollar panics is an extremely important question which probably has no good answer at the moment. Only in hindsight will the answer be possible.

Some analysts predict a fall in the value of the dollar of between 30% to 50% relative to the euro and oil. This of course will have a huge impact on the U.S. economy. Watch this issue carefully. Consider well the question of where to find the most credible sources of information on this topic. And don't fall for knee-jerk, old-fashioned answers, especially from Americans connected with the U.S. government.

The Power Of The Blank Buck

Here below is another way of seeing what is going on with dollars and euros. Often times, the best way to try to understand issues like money is to see it over and over again from other angles. Try this one, it is as simple as your checkbook. It was prepared by a gifted environmental writer from Australia who wants everyone to understand the euro.

> "Imagine this: you are deep in debt but every day you write checks for millions of dollars you don't have -- another luxury car, a holiday home at the beach, the world trip of a lifetime.
>
> Your checks should be worthless but they keep buying stuff because those checks you write never reach the bank! You have an agreement with the owners of one thing everyone wants, call it petrol/gas, that they will accept only your checks as payment. This means everyone must hoard your checks so they can buy petrol/gas. Since they have to keep a stock of your checks, they use them to buy other stuff too. You write a check to buy a TV, the TV shop owner swaps your check for petrol/gas, that seller buys some vegetables at the fruit shop, the fruiterer passes it on to buy bread, the baker buys some flour with it, and on it goes, round and round -- but never back to the bank.
>
> You have a debt on your books, but so long as your check never reaches the bank, you don't have to pay. In effect, you have received your TV free.
>
> This is the position the USA has enjoyed for 30 years -- it has been getting a free world trade ride for all that time. It has been receiving a huge subsidy from everyone else in the world. As it's debt has been

growing, it has printed more money (written more checks) to keep trading. No wonder it is an economic powerhouse!

Then one day, one petrol seller says he is going to accept another person's checks, a couple of others think that might be a good idea. If this spreads, people are going to stop hoarding your checks and they will come flying home to the bank. Since you don't have enough in the bank to cover all the checks, very nasty stuff is going to hit the fan!

The dollars the USA has printed, the 'checks' it has written, would start to fly home, stripping away the illusion of value behind them. The USA's real economic condition is about as bad as it could be; it is the most debt-ridden nation on earth, owing about US$12,000 for every single one of it's 280 million men, women and children. It is worse than the position of Indonesia when it imploded economically a few years ago, or more recently, that of Argentina."

copyright 2003 Geoffrey Heard, who is a Melbourne, Australia, writer on the environment, sustainability and human rights; gheard@surf.net.au.

Watching For The Panics

By this point it should not take much imagination to see that oil priced in euros is far more threatening to the U.S. than all of the magically disappearing weapons of mass destruction that Saddam purportedly was about to use on Americans. Seeing this, many Europeans argue quite heatedly that the "real" reason Bush invaded Iraq is oil. Who can say they are wrong? Bush, the heir to an oil fortune? Cheney, whose personal fortune is based on oil?

How can it not be about oil? Occupying Baghdad will ensure that Iraq returns to using dollars. An American-backed and chosen "democratic" junta in Iraq will enable the U.S. to pretty much ignore OPEC's production and price cartel for oil. Since Iraq has the second largest reserves of oil in the world and is capable, with additional investment, of pumping something like 7 million barrels of oil per day, Iraq is second only to Saudia Arabia as the world's most valuable piece of real estate. The U.S. can increase Iraqi oil production to levels well beyond OPEC quotas and drive prices down worldwide. Or, it can regulate the price of oil in U.S. dollars and shore up the pre-eminent value of the "petrodollar" for many more years regardless of the euro.

Here is how it worked for Iraq before the invasion. Iraq began selling its oil for euros in November 2000 when the euro was worth around 82 cents, It was seen by many at the time as a bizarre act of political defiance. But it was in actuality a shrewd financial decision. The euro gained almost 25% against the dollar during 2001. That was pure financial profit for the nation of Iraq.

Iran has hinted at adopting the euro for all oil sales and Venezuela, a large oil producer, is reputedly underway in 2003 with a switch to the euro. And so most likely is OPEC as a whole. At the OPEC meeting in Spain on April 14th, 2002, OPEC members began talking about pricing its oil grades in euro denominations.

U.S. controlled Iraqi oil will neutralize anything Iran does and Venezuela probably cannot make much difference in the world. But if all OPEC production went in terms of the euro, pressures on the dollar would begin. Unless... Unless the House of Saud, which reputedly owns something like a trillion dollars of U.S. equities and the world's largest oil reserves, maintains its prices in terms of dollars alongside Iraq.

Is this what the Imperial Faction has in mind? Is this what Osama Bin Laden really has in mind? Is Saudia Arabia, then, the ultimate target of Al Qaeda, most of which is controlled by radical Saudi's?

In May 2004, an additional 10 member nations will join the European Union. The Union will then collectively consume 33% more oil than the United States. In order to stabilize their mutual trade, the Europeans may increasingly pressure OPEC to trade in euros. Since the euro zone will be OPEC's largest customer with the most highly valued, stable currency in the world, this is a proposition which many of the OPEC nations will find attractive. If OPEC, or a portion of it, were to switch at that time to the euro for oil transactions, there will be a fairly quick change in relative values of the dollar and the euro, regardless of how much change has occurred between then and now.

If control of Saudi and Iraqi oil slips away from the hands of the Imperial Faction, which is entirely possible, the dollar would crash sooner than later. In essence, these oil-consuming nations would progressively flush dollars out of their central bank holdings to replace them with euros. Some analysts predict that the dollar could be forced downward by as much as 40 percent within a year if all of OPEC switched pricing into euros at nearly the same time.

Such a drop is likely to be catastrophic to the U.S. economy and would immediately begin to induce the collapse of 2006, perhaps two or three years in advance of the bottoming which will occur in 2007. Financial markets would erupt in massive reactions and the value of the dollar would spiral down rapidly. As the currency falls, there would be very little point for foreign investors to maintain their portfoliios of U.S. stocks, or, for that matter, any dollar equities such as real estate.

Since stocks values are already substantially over-priced in dollar terms, they could not hope to appreciate in value to regain the loss of the value of the dollar. Accordingly, foreign investors would have strong incentives to sell dollar-denominated stocks in favor of euro-denominated stocks.

How likely is a deep, sudden shift? It is likely to end up as a deep shift, but it is likely to occur over a period of two to three years. Even if control over the pricing of OPEC oil slips away into the euro zone, it is likely to be a gradual shift. No one could cope with a massive currency drop all at once. No one would win in such a situation. Governments and banks would simply freeze all accounts and transactions to enable them to more slowly sort out the demands and the needs. Even bitter enemies will cooperate to maintain some stability in the value of the dollars which their central banks hold as "reserves" for the World Bank system.

U.S. control of Iraqi and Saudi oil, kept firmly priced in U.S. dollars, likely will provide the political and economic force to fend off a lot of the estimated 40% drop in the value of the dollar which could develop. And even if Saudi Arabia fell into the hands of Osama Bin Laden, its new radical government would not willingly devastate the dollar and risk losing 50 to 90 percent of Saudi Arabia's accumulated world investment assets, which are valued as of 2003 in dollar terms at a trillion dollars plus. But all stabilizing factors in the international economy would vaporize if the Saudi Royals stupidly took the state's dollar assets with them as they hastily fled from an Islamic Jihad coupe. Under that circumstance, the radical Islamics of Al Qaeda (presumably) would most likely move to immediately destroy as much of the value of the dollar as they could.

The Shift List

Here is what to look for during the next few years:

1. Slow sinking of the value of the dollar vis a vis the euro; perhaps as much as 5%-10% by January 2004.
2. January - May 2004: Discussions and announcements by members of OPEC (Organization of Petroleum Exporting Companies) about euro pricing of oil.

3. January - May 2004: Surges in value of the euro in anticipation of the 10 new members of the eurozone and re-pricing of OPEC oil.
4. Increases in “China Wal-Mart Inc.” prices.
5. Decreases in U.S. Imports.
6. Expansion of U.S. Exports (may be in the form of weapons).
7. Announcements of expansion of the eurozone into more countries.

Of everything on this shift list, the most important is number two. How the euro plays at that time will tell us a lot about the next couple of years. For the remaining items, major changes in any of them tell you that the international economy is shifting rapidly and is probably acting to destabilize the dollar and the stock markets. These shifts may tend to occur in the order they are presented above, but they could swing in a different order. No matter how they occur, they are all interconnected and each one will be influencing the others. So continue to look across the whole range.

Section Seven

The Collapse of 2006

This book has built up an extensive dossier in the preceeding six sections about the political, social, media, and economic changes which are currently sweeping the world and which will drive the world's economy into a general melt-down by 2006 and 2007, if not earlier. Each Section is like a portion of a three dimensional hologram. In this final Section, the sections of the hologram are shown in a unified composite to present, without much qualification and analysis, selected images of some aspects of a likely future. Without the explanations and context setting of the prior Sections, this future in this Section moves fast. Are you ready?

Timing Of The Coming Collapse

As discovered from hints given by Edgar Cayce during the 1930's, it is now easy to observe that every 24/25 years a major economic realignment and depression or deep recession recurs. This cycle does not include all recessions, but, for as far back as economic statistics can be found, this cycle does rigorously define the exact occurrence of all the worst depressions and recessions in the United States. (See "*Return of the Phoenix Book Two: The Great Break-Up*", keyword "economic cycle")

This cycle has the potential to inflict economic crisis as bad as the Reagan recession during 1982, or even as bad as the Great Depression which bottomed and began to turn in 1933 after three years of slumping. Since the "***structural problems***" in the American economy are so many times worse now than in 1982, even apparently as much if not more so than in 1933, the most appropriate comparison is likely to be the Great Depression which ignited the FDR "New Deal" reforms during the 1930's. Or, one can compare coming events with the Great Depressions of the 19th century which ignited the populist and progressive movements and drove the eventual election of Teddy Roosevelt and his aggressive moves against the great "Trusts" of the wealthy elites.

As for fixing the "***structural problem***", discussed in Section Six, this is a matter of focusing on the "***giant sucking sound***" which surrounds acute job loss. The ***giant sucking sound*** was first heard during the 1980's and the din has gotten progressively louder through the years, louder and louder through nearly the length of this entire 25 year cycle. It has created a gaping wound in the economy of North America of such size that many Bear analysts suppose that a collapse into the required economic and political adjustments will be similar in magnitude to the dislocations of the Great Depression.

This "***structural problem***" is elementary and not arcanely hard to understand. No matter what type of conventional stimulation is applied, it appears that job growth will not occur in significant numbers any longer in North America, the stimulation for the American economy is merely exported to China right through gaping economic holes (U.S. "global" corporations). More stimulation and more "turn-over" expansion simply make the giant sucking sound even louder. Lots of goods cross the docks. More jobs get created in China, more jobs disappear in the U.S.

Eventually such a system must collapse in on itself by shrinking into a steeper and steeper depression. Thus eventually all the bubble equity and salary valuations made in the U.S. also must crash into the ground. That is how God recycles the nations in the great cosmic cycles of history.

The emergence of this type of structural problem has been forecasted now for over a dozen years by those with anti-globalism attitudes, including Pat Buchanan, Ross Perot, and many among the Green Party circles. If the economy does not expand in North America as a result of this year's war stimuli, it may very well be that China has become a huge Wallie World even for Pentagon buying agents.

Most conservative analysts of the American economy and the stock markets tend to agree that the DJI stocks, with a price earning ratio of about 30 during May 2003 (when the DJI was at about 8500) are overpriced by about 100%. This means that, despite the drop of the New York stock market from a DJI average which peaked at slightly above 11,000 in 2000, stock valuation is still on a bubble, twice as expensive as it should be during May 2003. Along this same line, many analysts point out that most NASDAQ technology stocks have bubble valuations which are considerably more inflated than this.

From this simple assessment comes the basic Bear position. The values of stocks (as of June 2003) are more likely to stay as they are. Stocks will not gain much in value until economic growth and corporated earnings pick up much further during the next few years to reduce the price/earnings ratio back into the healthy range of around 15:1.

So, for Bears, stocks in general are not a worthy investment for at least the next couple of years, better to buy things like real estate, euros, bonds, or maybe, just maybe, new technology companies which are already profitable but still very small. Small investments in such companies have room to grow like Microsoft did and only a little capital needs to be at risk to obtain potentially huge gains.

The more sophisticated Bear position analyzes the logic of the structural issues of international trade, the petrodollar, the economic conditions in Europe, and so forth. Bears try to fathom where to be invested and how, if at all.

The question, IF AT ALL, is a profoundly important one. There are times in history to not be invested in anything. Given the unsettled and polarized atmosphere in the world today, which is causing the huge political, economic, and media global shifts as discussed in the previous sections, it is highly likely that the Bear position will continue to find a lot more negatives than it likes during the next few years.

Keep in mind, especially, that the Wall Street Collapse of 1929 preceded the bottom of the Great Depression by approximately 36 months. It took a full two years for stock values to erode away to $.10 on the dollar. It wasn't until 1932/33 that tens of millions of people were jobless and totally broke.

By this comparison, one could argue that the stock market of today is already in the collapse mode similar to the early 1930's. One can argue rationally that the economy will simply decline away to nearly nothing to objectively prove to Americans what is already perfectly obvious except to the perfectly obdurate materialists and wealthy elites for whom money is merely large numbers, namely that they no longer have an economic system which works for people and far too much real production of real goods has been transferred overseas.

However, conditions are radically different than in 1929 and in fact today the entire world generates the demand in the American equity markets. The Saudi's, for instance, own something like a trillion dollars of it. The world has become accustomed to this and it will take considerable time for confidence to fully erode and thus cause investors to replace American investments with Eurasian investments.

In sum, the Bears are essentially right. A lot of major adjustments are overdue. It is just a matter of timing. Accordingly, do not expect the Bear position to disappear, even if the stock markets show a little lift. Bear signals will continue to be given and widely discussed, and this will continually confuse the Bulls, making them considerably skittish. This of course will make a serious new Bull run more difficult to sustain than not. The net result is likely to be a market rise which is full of fits and restarts.

Advent

The Bears may be right but they have no monopoly on the truth or the auguries. Bulls are also correct at times and even now they are correct at least to a limited degree. Here is why. Many corporate elites and investment management circles in the U.S. are going to exert an enormous amount of effort to ignore the obvious (structural) issues and, like Bush and the Imperial Faction, tell a lot of tall tales while they attempt to huff and puff a recovery into the economy. After all, a lot of Boomer and Gen. X net worth, including those who are active in the main stock trading companies giving the signals, is based on stock options which will need a huffed and puffed stock market to provide cushy lifestyles with multiple residences, private flying, and all that. So expect very valiant efforts by the wannabe's and nearlyso's to line up crowing clubs among the New York Media for the new next new wave of puff.

Given the truckload of signed blank checks the Republicrats in Congress have given Bush and the Imperial Faction for massive new expenditures in the military-industrial complex, it is likely the Bulls will have some success in generating an increased turnover of goods and services to give the appearance of a recovery. This may succeed in causing a slow forming bubble in the DJI which may peak out in about 2005 or perhaps as late as 2006 at something around 10,000, perhaps even flirting with 11,000.

Thus, even with the terrible polarization being induced by the Bush administration at the current time, U.S. equity could modestly appreciate again during the next few years, enough even to generate another (short) bull run before the cyclical Wall is hit. But, this recovery probably will be problematic – dicey, incomplete, spotty - it is already being called on the street "***a jobless recovery***" in recognition of the huge impact the "***giant sucking sound***" is having.

This will mean that personal incomes will not increase much in the U.S. and that discretionary spending will likely continue to shrink.

That means that a great many consumer industries, especially in the high tech area, will stagnate at breakeven levels. People just don't need another computer upgrade, nobody "needs" to buy another Microsoft product, nobody needs another TV, phone, etc. And if you do need any of these things, the classified ads and Goodwill are excellent sources for them at pennies on the dollar of the original prices.

Accordingly, look to the next 30 months as a period of basic stagnation in many things, the same sort of economic malaise which set in during the Carter years just prior to the Reagan recession in 1982. Keep in mind that some regions and some economic sectors will do better than others. Housing construction, for instance, will do even better in the Pacific Southwest, in fact this industry in some portions of Arizona never really experienced the slowdown of the past three years. Retirement moves are continuing a considerable growth in many other regions as well, especially in rural areas.

Look also to what is happening abroad, especially through the new internet-news media. Look for rumors and signs of what is happening to the world's basic economics, especially in regard to the primary "shift list" variables which are discussed in Section Six. Is the petrodollar unraveling or no? The magnitude of your problems will be directly predictable from the magnitude of the changes in the items on the list.

Above all, keep a highly skeptical mind about anything said from anybody connected with New York based media or brokerage firms or the Republican Party, or even by the faux-centrist Democrats for that matter. Many of them just don't get it and will continue to attempt to sell you obsolete ideas about what you should do. They will ply you with knee-jerk responses which are designed to create more wealth for the elites, devil take the hindmost.

Those who create the crazy "Stupid White Man Culture", as Michael Moore is fond of calling it, will talk knowingly in terms of control of markets, dominance of nations, manipulation of markets, power, gaining the upper hand, winning through greater competitiveness, cutting costs, and so on. These are now all merely code words denoting obtaining more and more privilege for less and less people. The end result of that logic is the King On The Hill surrounded by the plains of universal bankruptcy on which revolutionaries are beginning to organize and pass the ammunition.

During this stagnating period, during which expansion rates are not likely to exceed 2 percent annually, the political world will continue to fragment in North America even as it consolidates in Eurasia. Electoral politics in the U.S. during the elections of 2004 will be full of conflict, disputation and increasing fragmentation over practically everything, ranging from the existence of the Imperial Faction, terrorism, the invasion and occupation of Iraq, the economy which needs Viagra, the loss of jobs, the incredibly shrinking middle class, the alienation of Europe, etc.

Meanwhile, Eurasia will increasingly pull together, strengthening its economic union. This will be accelerated by their stand-off with the Americans over how to use the United Nations and the proper methods for creating peace in the Middle East, and an extreme personal dislike för George Bush. A sliding dollar will add considerable anxiety into the mix. Europeans may need to create an integrated economic stimulus program for their own economies and adapt ways to cope with their loss of exports to the U.S. (which could get very expensive very fast during 2004 if the U.S. dollar declines as discussed in Section Six by 10%-40% from it value in June 2003).

The Political Struggle For Recovery

Politics will zig and zag tremendously during the next several years. The sudden changes and dramatic events will create great controversy, passion, possibly even occasional violence from over-stimulated police and demonstrators. A great deal of the controversies will be over fake arguments, bogus information, big lines, big whoppers, media slants, vicious tactics, personal slandering, and 'attitude". Best not to get caught up in the bogus discussions.

During the remainder of 2003 and 2004, a great part of the controversies will be about the invasion of Iraq, the war on terrorism, the proposed war on Iran, the increasingly questionable competence of Bush, Rumsfield, Powell, and Rice, and not least, the emergence of the "Imperial Faction", which is seemingly a narrow ideological group who have highjacked a large nation and who now seek to convert it into their personal hip-pocket empire.

One should expect the concern over the "American Empire" and American Imperialism to become more and more vocally expressed in various forms throughout North America, especially in various regional and local media, and, naturally, in political controversies and debates surrounding the upcoming elections of 2004. More and more, questions will be raised about whose interests are really being served by American militarism. Factionalism around the issues of Palestine and Zionism will become more and more prominent.

The peace and social justice movements of populism within the U.S. are rapidly waking up to the New Wave Jingoism which is being sponsored by "Billy The Kristol and The New American Century" gang. A very grumpy liberal left in American politics is already proactively engaged in preparing to wage a sustained battle against this new generation of "neo con" fascists. The left is very tired of quiche feel-good bubbles and Clintonesque moderation. They are ready for some seriously aggressive meat-eating.

As the months roll on, the economic issues will become ever more pressing and this will increase the populist voltage considerably. All things considered, events disfavor centrist, modest positions proffered by those who are primarily seeking to maintain their cushy careers. Events increasingly will favor strong opinions which are directed in support of what has been described as populist positions. Along with this will come a strong resurgence of the "class warfare" rhetoric which makes Republicans so livid.

Unfortunately for those few remaining sensible, principled Republicans of the old "Eisenhower-Dirkson" school, the very grossness of class corruption in America has such an unmistakable stench the length and breadth of the land, they shall not be able to box it this side of the collapse of 2006. A huge emotional wave of reaction comes. Candidates will arise who will accuse the Republicans of personally betraying tens of millions of people. The basic humanhood of Republicans will be called harshly into question and principled conservatives of the old school will be forced to confront even in their own party a lot of things they hate .

Initially, the main casualties of this emerging political bloodbath are likely to be the faux-centrists (such as Gephardt, Lieberman, and Biden) who have dominated the Democratic Party since the Carter years. They sound and look far too much like Republicans to appeal any more to the left. They are too easily suspected of being fellow travelers with the Imperial Faction and indeed their timidity with Bush's gross distortions and loopy, neurotic, juvenile handling of foreign affairs places them by default into a supporting role for the New Wave Jingoism.

Accordingly, where ever faux-centrist Democrats control political offices, the political left is more likely than ever to bolt the Democratic party during the ensuing years in favor of alternative "non-establishment" candidates. This will result in Republicans winning races they should lose but in the longer run this will effectively help dissolve faux-centrist careerism in the Democratic party. In the long run then, collapse of portions of the Democratic party will eventually increase support for a more assertive movement on the left.

The faux-centrists will of course use the traditional "Plastic Argument", that it is better for a plastic Democrat to "win" an election in order to prevent the Republicans from winning a vast majority of seats in the Congress. But the total corruption of the electoral process by money during the past 20 years has utterly destroyed this argument. As things now stand, all you ever get by voting for plastic Democrats are plastic careerists whose timidity stands in the way of vital leadership on principles and just causes.

Twelve years of plastic Democrats and we still don't even have a health care plan which at least Democrats agree on. Voters and rank and file must ask: "Isn't it utterly pointless to continue down the road in this company?"

The failure of the plastic strategy will shock the faux-centrist portion of the Democratic Party. They will argue vociferously this next year that even more plastic is needed. Senator Lieberman, for instance, is arguing that the Democrats need to give the appearance of having morphed into an Imperial Faction George Bullyboy look-alike on international affairs. Poor soul, Lieberman must be still lost in the shock of 911. Time has already passed 911 by and it is now the Bush Bullyboy imperial visage which is one of the main reasons why political support is shifting away from the Republicans while the so-called political center dissolves.

Enough people are waking up to the dismal historical results of plastic faux-centrism that the Plastic Argument will no longer work nearly as well as it did. Now motivated by the amazing polarizing ways of Bush and the Imperial Faction, democratic partisans of the liberal left will come out in droves to turn American politics upside-down. The left is already signaling heavily that it is determined to have a complete remake of the Democratic Party, elections tailor-made for careerists be damned.

Not to be outdone on the left, the traditional conservative base of the Republican Party is waking up to the overweening ambitions of what they have been calling the neo-conservative movement. Voices from the old-line conservative right are now beginning to call for outright domestic political warfare with this "Imperial Faction" and their supporting cast of "evangelist" born-again choirboys. Recently Pat Buchanan weighed in on the right with a powerful synopsis of the activities of the Imperial Faction. He called for Republicans to throw them out of government or face dissolution of the party. He also called for George Bush to disavow the PNAC plan publicly.

From this it should be obvious that Republican politics may be as interesting during 2004 as Democratic politics are certain to be. But the corporated elites who sponsor Bush will likely contain the Republicans in 2004. Thus the first casualty of the growing polarization in the United States probably will be the Democratic Party.

The Democratic Party has morphed a couple of times since Roosevelt and is about to morph again. This so-called party of the "common man" became corrupted after Carter into a "centrist" party dominated by careerists who serve mainly corporate secular interests while serving up only lukewarm leftovers and crumbs for the social agenda of the past (the old FDR coalition) .

Under the faux-centrist careerists, the rank and file of the Democratic Party has endured nearly complete capitulation of the Democratic Party during the past 20 years to the natural predators of the common interest, first to the corporate mindset and now most recently to the Imperial Faction. The ORGANIZED, DELIBERATE, CONSCIOUS silence of the Democratic Party to the usurpations of first the corporate "globalists" and now the Imperial Faction fully reveals the extent of this corruption by careerists.

Currently, a core of less than one half of the current "Democratic" members of Congress sense the higher values and standards of conduct sufficiently to require that American affairs be conducted legally in ways which satisfy both the Constitution and international law. The majority of Democratic office-holders this past year have given the clear odor that they support the continued usurpation of the Constitution and the power of the Congress, as well as flagrant violation of the United Nations Charter and many international laws and treaties to which the U.S. is legally bound. Many will continue to assert that it is better for Democrats to tiptoe around and not rock the boat very much about the niceties of these legalities.

Given the huge influence of money in buying elections, the odds are currently heavily stacked against freeing the Democratic Party from its corporate sponsor-owners and those who prefer to "tiptoe" silently through the Halls of Power on their way to the cushy Retirement Room. Accordingly, it may take much more severe economic pain to motivate enough change over in the party to make it a suitable vehicle for majority rule by an egalitarian class.

Nevertheless, with or without suitable pain to motivate serious change, it is likely that a pitched battle will be fought within the Democratic party during the next two years for its heart and soul. If ways can be found to change the equations in the game, the party may be able to regain coherence around a National Economic Industrial Policy with a big heart. If it is truly engaging enough, it may be able to re-forge a coalition of power which can turn the wheels of governance under new management toward a better direction.

But it is unclear as of July 2003 that social reform forces will flow in proper directions in sufficient quantity to sufficiently re-forge the Democratic coalition. The heavy presence of several faux-centrists who have announced their candidacies for the Democratic presidential nomination argues against any major shift in the Democratic Party during 2003 and 2004. The faux-centrists, who will tend to awkwardly goon around as Clinton heirs, will instinctively use a lot of corporate cash to paint tall tales about Republicrat nostrums while glossing over heart-breaking class issues. Those who will be talking about restructuring the party and its strategies and platforms in ways more relevant for real people will have a hard time being heard above the din of confusion emitted by faux-centrist Republicrat bucks.

For voters seeking real (progressive populist) candidates amid the confusing din of claims and counter-claims, the task will be made relatively easy by the Mass Media. In its traditionally bigoted manner, the Mass Media will tend to "black list" real reformers as "unknown", "controversial", "dark horses", "with little support", who likely have "little chance to win". This is the standard "slant" line which is used by the media to chase away real people from the electoral process and announce that they are not really a legitimate part of the careerist system.

These labels are in fact part of the schmearing process which guarantees elections only to those people who are well enough connected to get backed by big bucks who are going to be spending lots of money on media advertising. Look for those "black list" labels, then you know whose campaign literature to read. Those candidates will eventually win. If not in 2004, then in 2006 and in 2008. Start voting for them early on, and often, so they will be encouraged to run when the pain is the greatest.

It is highly likely that great energy will flow into attempts to kindle the third party movement. There well may come new parties on both the sorta-left (the Greens) and the sorta-right (Conservatives). This flow may upstage efforts to reform the Democrats and this may deny the Democrats an ability to muster majority votes in many elections. Accordingly, the Republicans may continue to hold substantial power, including the White House, up until at least 2008.

More than likely, the American social fabric will progressively unravel under continued Republicrat sponsorship of increasingly fascist, would-be imperial rule. As well, things could turn very sour internationally during the next several years. Setbacks and strained, paralyzed entanglements could, almost certainly would, substantially increase the polarization in North America.

Continued belittlement of Iran by U.S. chicken hawks could stimulate Iranian factions to support guerilla resistance of the U.S. occupation of Iraq. The occupation of Iraq could turn into incessant violence as in Palestine. Afghanistan could continue to sink into anarchy and eventually require a much larger international force to occupy it, as during the Soviet era.

Al Qaeda might make good on its claim to overturn the Saudi monarchy and replace it with a militant Islamic Republic with extreme attitude towards the U.S. Or, failing that in the short run, radical "Muslim" fundamentalists may engulf Saudi Arabia in an extremely violent uprising. American involvement in any of these long and costly entanglements will mainly result in the social fabric of North America unraveling at a faster rate in a more disruptive, bitter, and violent manner.

If Americans can genuinely re-forge a majority vote for a National Economic Industrial Policy with major heart (which is actually supported by its office holders) during the 2004 election process, the decline in American equities, values, and effective wage rates (relative to other currencies) may not fall as much as during the last Great Depression. Unfortunately, Americans may not have had enough time to fully realize that the levee is dry and that they must take a big hike into a new world.

In this process of re-alignment, the Democratic Party will either fold into a mere pittance or it will return to FDR-style policies to serve a renewed "New Deal" coalition. But it matters not at all who implements such a strategy. It does not matter a whit if this policy is promulgated under Democratic or third party auspices.

In high probability, it is likely to be promoted initially under third party auspices. Regardless of how, it comes. A new populism comes. A sudden, pervasive shift in the rule of government under populist leaders comes no later than 2008.

If the process of this transformation requires four years of grinding privation between 2004 and 2008 to make the lesson clear to enough people, so it will be. If so, a new FDR "New Deal" coalition will not re-forge sufficient policy and sufficient party to assume power until 2008.

The candidacies of Kucinich or Dean are excellent bellwethers for the condition and direction of the Democrats along this populist road. If Democrats connect well with these two national candidates for the Presidency, the Democratic party may realign as the new progressive populist movement. If the majority faction of Democrats remain mired in their faux-centrism affiliations, the Greens are likely to emerge as the primary agent of change.

The New Deal & The National Economic Industrial Policy

The last time the U.S. was technically collectively bankrupt (literally) in 1932/33, the FDR-coalition allowed the wealthy to eventually recover their corporate paper positions. A great many large family-dominated corporations and banks were given easy access to credit and government contracts to insure their survival in return for a vast and rigorous explosion of social policies to create "a new deal" for the common man.

This was apparently the best compromise which could be made at the time. It was a generous compromise but it has been the undoing of the entire last 20 years. It is not likely that the holders of paper equity will be nearly so lucky this next time around. There are going to be a lot of revenge agendas floating around with a much more acute sense of history. The price of salvation for paper equity is going to get a lot higher.

Some of those wealthy families who survived the Great Depression, whom FDR and the Democrats directly helped save, have been the main sponsors of right wing causes and movements. They like the easy money, the easy credit, and the vast government sheltered contracts, but they can't stand making room for the common man. From some of these families and movements have sprung up the modern groups who are hell-bent on smashing "New Deal" social policy and transforming the American government into a tool of old-fashioned imperialism, akin to the era of jingoism during 1898-1929.

They have succeeded to a remarkable extent. During the 1950's and 1960's they gave rise to the John Birch society and similar paranoid lunacies by which they made the conflict with communism as bitter and as violent as possible. The current queer impasse with Cuba is one of the enduring fruits of their nutty thinking.

Under Reagan's reign, they began the process of "deregulation" by trashing the "New Deal" social economic industrial policies which created the post war wave of prosperity. In the name of innovation and progress, Republicrats turned the electrical power utilities and telecommunication companies into a vast universe of semi-privatized confusion and telemarketing scams, many of which are now heavily burdened with huge piles of New York debt, bankruptcy proceedings, and State Attorney General Consumer Fraud lawsuits which are seeking untold billions of dollars in compensation for defrauded consumers.

On top of greed-mongering deregulation, right wing partisans added the creation of a network of right wing institutions, think tanks, and national publications to publish and promote corporate "our thing" agendas. They chased this feast with a major buy-up and consolidation of mass media control by which they could ply their nostrums into the hypnotizing dimensions of Mass Sorcery. Thus girded and armed, the right wing eventually captured enough of the Republicrats to elect George Bush Junior.

Democrats Stop Being Stupid And Get This!

With these same institutions and tools, the right wing authored and refined the Bullyboy Strategy during the past 12 years for dealing with liberals and Democrats with negative attack ads and lies. They succeeded in eliminating many viable candidates with a wall of garbage. These campaigns were so successful in trashing Democrats and intimidating the left, there was nothing to stop the Imperial Faction from carbon- copying the strategy for dealing with Arabs. So they did, most evidently. They easily used the same slandering negative attack tactics which they used to neutralize Clinton's presidency to advocate the invasion of Iraq to destroy Saddam. They engineered a brilliantly successful Mass Broadcast Media promotion campaign to procure just enough political support to get away with it. With two regime changes down, in May 2003 they began using the same tactics to start the ball rolling over Iran.

From this we can see that it is truly remarkable what can be accomplished when you invest serious money in private efforts to transform the politics of a nation. Ironically the right wing funders of nominally "conservative" institutes, groups, and media, who haughtily abhor social policies for national salvation, have proven beyond doubt that one can in fact, in contradiction to right wing claims, effectively induce a major change in a nation. If only they could recover their senses and regain their heart to do something decent for people.

The Cayce Recovery Strategy

Given the huge scale of the "***structural problem***", i.e., the lack of an actual economy and logic of organization which works in the broad interest of people who do the work, it will be immensely difficult for a large movement of people to reason out a coherent plan which is comprehensive and aggressive enough to stand a chance of working to solve the "problem" (the giant sucking sound).

The history of previous eras of major change shows that it takes many years to fully implement profound reform. Thus, one should plan on many years of politicking before the next era of populist reform is implemented in ways which meaningfully impact one's life. But, regardless of the exact time dimension, eventually the U.S. will end up adapting an industrial development strategy to provide the dynamics of growth for the next 25 year cycle (which will end in about 2030/2031). Hopefully this new ladder of economic upliftment and personal growth will be available soon enough to be helpful for all of the people who read these words.

In the meantime, how can one best position one's personal situation and family affairs to survive the next several years of stagnation, collapse, mass privation, and gradual recovery in the new era? The positioning on a personal scale is best conceived to take two forms. One is to be better connected to real resources in areas which can directly supply real water and real food. The second form is to prepare to become active as a worker, professional and/or investor in the most dynamic fields of endeavor, in industries which are most likely to be the primary engines of growth during the next historical expansion wave. For large numbers of people, this means quitting the old loser industries and picking the new winner industries.

The last part of this Section explores some of the most likely new industries and the most likely losers. But before these are taken up, one can learn a great deal about the lessons Edgar Cayce gave during 1929 and early 1930's about how to recover from the Great Depression. During Cayce's time, the worldwide economic collapse after 1929 was generally conceived as nothing less than the general collapse of civilization. His lessons provide a crystal clear logic for how to approach the next collapse and recovery.

Just prior to the crash in October 1929, Cayce outlined a masterful investment strategy for his client's future, which encompassed more or less the second half of the 20th century. It is a simple strategy in its basics, and easy enough for any one at any level of the economy, whether as an investor, or as a professional or worker. Cayce began by pointing to the fundamental changes in the nature of the economy. He suggested liquidating investment out of the old industries, those which had been dominant for the past 70 years. He strongly advocated that his clients get into the new industries of the future.

What did this mean for Cayce's clients? For the era of his day, this meant getting away from investment in the railroads (which were the dominant blue chip stocks during the last half of the 19th century and the early years of the 20th). Liquidate these, Cayce told his clients, and put your money into the new field of mass communication, anything of the "ariels" as he quaintly described the new industries of radio, telephone, telegraph, and aviation.

In May 1930, despite the deepening depression, Cayce predicted the industries which would become the next U.S. focus of industrial activity: ***"those of the aerial nature; that is, radio and its associations; aerial and its associations".*** At the time, both the radio and aviation industries were still widely regarded as commercially unproven toys.

Cayce suggested that stocks could be bought in these fields AFTER the crash of 1929, but only on the basis of conservative price-earnings ratios. These would be BOTH the safest, if bought correctly with no speculation bubble in the stock prices, AND the very best speculative bets, because these new industries would grow the fastest. In other words, if bought at the beginning or "ground floor" of these industries, AFTER THE 25 YEAR BOTTOM IN THE ECONOMIC DEPRESSION CYCLE, one could in the long term profit handsomely without relying on speculator bubbles and rigged pricing.

Cayce foresaw that exceptional changes and improvements would come in the communications and broadcasting industries. Through the slow recovery period, he continually pointed years in advance to the coming of TV, fax, telex, expanded stock ticker-tape services, and microwave transmission on land to replace copper wire. All of these Cayce continued to tell his clients to purchase on the basis of conservative price/earnings values.

In February 1933, he named the ***"most outstanding change that would come ... aeroplane manufacturing and transportation industry".*** In November 1942, Cayce predicted that ***"air transportation, - will become more and more the basis of ALL relationships with other nations, countries, as well as the internal or national activity".*** By 1944 he predicted that ***"any town that has any name will eventually be on an airline and when they're off it's as bad as being off the railroad, (in another ten to twenty years)".***

Cayce also predicted in April 1936 that America would become what Roosevelt later termed "the arsenal of democracy": ***"America must remain - as it were - the balance of power in not only the money forces of the world but*** also in ***those influences that will later be for the manning of those ... greater destructive forces ... for manning or powering other nations with implements of destruction of every***

nature, these will show for a greater and greater volume being carried on in the United States".

In other words, Cayce told his clients to look to the industries on which the governments of the world are spending more and more of their money. Since World War II was looming, which Cayce had already predicted under the name "***the world on fire***", investments in companies which made munitions were bound to do well for several years. In November 1941, even before the U.S. had entered World War II, Cayce predicted that the peace would return in 1945-46 and that the prime focus of the post-war economy would be the consumer: "***the supplying of properties or activities to the GENERAL public ... there will be more and more demand for utilities, for household necessities".*** In June 1944, in his last economic prediction, Cayce signaled that great economic changes and expansion would come after the war ended, but it would begin after two or three years of economic disturbances. And in fact, it was so. The civilian economy began to bloom in 1948.

In September 1942 Cayce advised investing in "***Any electrical appliances, all forms of home appliances, those of greater conveniences, - all of these will offer the greater investment now and for the future - for the next few years, at least".*** These consumer products drove the economy for a long time and companies which manufactured them become huge. It is still true today, in the form of computers and digital electronics. In March 1944 Cayce again predicted post-war economic leadership for technology in "***all of the industrial products, including radio, ice box, automobile, lights".*** He predicted at the same time that TV would become an important consumer product before the TV broadcasting industry had even been conceived.

(For citations to Cayce's comments, see the Chapter Notes at the end of this Report)

The 2003-2008 Recovery Strategy

To evolve your own positioning, begin with the Cayce Recovery Strategy. Obviously you cannot copy the specifics. You can only copy the logic. This logic will work for you whether you are looking for a job, seeking to create a business, desire to reinvest your private wealth or intend to rapidly expand a new economy with a new National Economic Industrial Policy. After the next collapse, what will make sense? What is dead, old history and will not recover very much very fast? What is hot, what are the industries which government must back, what are the industries which will receive the most growth from new consumer spending?

We can also ask what positioning strategy makes sense during 2003 and 2004 to wring as much as possible out of what ever prosperity is left in this cycle.

Stocks - General

The Cayce Recovery Strategy boils down to common sense strict attention of the perfectly obvious. Out with the old, in with the new, but only if it is profitable and priced at a conservative price-earnings ratio at the (relative) bottom of the economy or at the relative bottom of growth for the industry the company is in. It is difficult to go wrong with such a strategy, if one knows how to intelligently avoid dogs while covering many bets.

Obviously the Cayce Recovery Strategy needs an entirely new slate of candidate industries for this first decade of Century 21. TV, Radio, and Aviation, such as we have known, are definitely not going anywhere... Or are they? In China, they are just getting started (but that's another story for a different people).

Positioning can begin now, but keep in mind the prime injunction. Buy equities which are conservatively priced at price/earning multiples in the range of 10-15. Keep in mind that a ratio of 15 is not "a bargain" as the fast crowd will tell you. Anything past 15 has froth in it and is on a bubble curve. A ratio of 15 for a stable, slow growing industry is about as high as it is worth. A ratio of 10 is worthless if the company is barely making any money and might go bankrupt next month. In other words, you have to know your industry and your company and from such knowledge figure it out on your own. The only exception may be "toots" who have the same ideas in mind that you have and who happen to know more about the industry. Since you are not likely to meet them very often, unless you happen to be rich enough to hire such toots, you won't be able to use them very often.

Many of those who preach in the nouveau economy of the high flying 1990's will tell you these figures are awfully conservative, that things have changed. It is true that these numbers are very conservative. That's why they will work. And things are still a-changing, away from the nouveau and right back to the old basics. Keep in mind that brokers are always unable to tell you precisely when the market is going to break down, which tells you how much they really know. They also never can tell you how much you will profit on a given stock....which means all of their statistics, in the bottom line, mean not much at all.

The most important moment is when you buy or sell. It is very simple for the next three years. For the general market, forget short term speculations. Growth will be too slow to yield much and the skittish ups and downs will drive you crazy. Forget speculation.

For buying, 2003 to 2006 is a lousy time unless you really are an expert at working the right kind of industry with inside information. Or, unless you know how to pick a winner out of the Winner's List. In any case, buy no bubble value. Buy only at PE of 15 or less for the long term. Hold through the collapse, make no further decisions about it until 2008 unless events force your hand. At any time, if you can buy a good candidate among an obviously winner industry at the correct PE ratio, do so. Obviously the best time to buy will be during 2007/2008. It will be hard at that time to find stocks with inflated values. Make sure you buy only known candidates from the winner lists which don't have a million lawsuits chasing them.

For selling, try to hold through to 2005 or at the moment the DJI hits around 10,000. If the DJI market shows no improvement by 2005, liquidate all stocks priced at PE of 16 or better at a suitable time, probably after the State of the Union Address. If the economy and the DJI market are improving, wait another year unless you can make more money some other way. Don't wait past the Summer Solstice. If the DJI market slumps through to February 2004, why wait? Sell all those with a PE of 16 or better. Or, the moment the DJI hits 10,000, sell. The most you could hope for is about another dime on the dollar but that is a highly speculative bet. Waiting for that is not worth losing fifty cents on the dollar, or more, when the Wall is hit.

Remember, the last people to know will be the Bulls and brokers on Wall Street. Their information will be worthless. Keep only those stocks that belong in the new era, in the new National Economic Industrial Policy. Be out of everything else. And try not to be in bonds. There are likely to be massive bond defaults. Be in solid real equities at conservative valuations, have everything else in reasonably secured liquid form, like government insured savings accounts.

The following lists are suggestive only. These lists are not exhaustive nor thorough. These Loser and Winner lists are just a starter for looking at some of the more obvious industries and for asking the Bear and Bull questions which need to be asked. To find the insider "hot growth" potentials, you have to pick an industry or field and become your own expert on its prospects. The world is an awfully big place and one author simply cannot spy and spiel out all the goods and industries which are equally suitable across continents and levels of interest. But perusing these "starter" lists may help you clue into the areas you want to explore on your own.

Businesses and Jobs

Out with the old and in with the new. In all cases, use your own superior sense of judgement about the particulars of your associations, career, and investments. Pick your industry knowingly, full as much real information about it as you can. Pick it because you like it and want to be an insider. Know the score, the tech, the players, the markets, etc. Play the field within the industry. Don't try to play the field through all industries, you will never be anything more than a bozo.

Dump The Losers

Let's start by getting rid of the losers.

In General

Keep in mind that some of these losers can be picked up after the Super Collapse in 2007. At revalued pricing, some of them may be worthwhile. Utilities for instance, will be big losers. They can likely be re-bought at \$.10 on today's dollars.

Conglomerates will be massacred. They are too big for anyone to know how to manage them. During the crash they will fall apart. Major Divisions may be buyable for a song and a tap dance.

Cayce's "arials" in America are super saturated. These can be bought at \$.10 on today's dollars after the crash. These include mass media, dot-coms, TV stations and channels, radio networks. Anything supported by advertising dollars will fall on very hard times.

Software and computer companies will also fall on hard times. Many more will disappear.

Automobiles – Traditional

The automobiles of the past few years, especially the crop of 2002 and 2003, look like a full employment program is underway to make sure that every wannabe designer has a job. There are more designs than buyers, and an awful lot of them seem like weird looking spaced-out wannabe Star Wars shuttles. No wonder there is a recession going on. People have to be crazy to buy a lot of the current crop of gas guzzling "east coaster" models for paranoid yuppies who work for the military industrial complex. It is time for a long break while car manufacturers get back to basics.

Investments, jobs, contracts, and business in most aspects of the auto industry are just going to get riskier and riskier while these companies flop around trying to rediscover common sense.

Aviation Civil

American civil aviation is dead, dying, or going down down down. Too much terrorism, hysteria, cheapskates, compulsions, and government goonery are killing this industry. It will take a new political era in the world before this industry makes a comeback. When it does, flying will be a lot more expensive and it will not operate like the "sardine" industry of the past 50 years.

Bioengineering & Pharmaceuticals

These industries are now driven by Dr. Strangelove look-alikes who spend billions of dollars to lure you into getting hooked on "maintenance solutions" which you will consume forever. Imagine about 10,000 Dr. Strangeloves deviously inventing ways to outfox everyone else to control a major chunk of the economy and the Earth itself. This will give you a pretty good idea about the mentality in this industry.

Sure, occasionally there are some gems here. But most bioengineering products will fail. With each passing year, the negative reactions mount larger and larger and the probability of eco-disaster increases. Were it not for the collapse, this industry would eventually be busted out by trillions of dollars in law suits and by political pogroms waged by really pissed-off natives.

The essential problem of this industry is incurable. Dr. Strangelove wants to engineer radical new biochemistry for mass applications under patent controls. It is not about health. It is about patent control. Repeat this three times so you have this embedded strongly enough to prevent mind control erasure by the Mass Broadcast Sorcerers.

The incurable part of this is that the validity of the science of these products cannot be really validated in anything less than a generation. Who knows what kinds of dementia or degeneration will onset from this stuff unless the impact an the actual aging process is observed over a period of decades? Nobody knows. Nobody cares. Dr. Strangelove is in a hurry, he wants his billion dollar stock option sale in about three years, not 30.

The big "drug" money-makers of this industry are going to fail. We already know enough from history to know that the odds are high that some portion of this stuff is going to produce queer results in the long run. Billions and billions and billions of lawsuits. But there is much more here than that. A lot of these new medicines are simply a rip-off. Cheap, traditional, herbal-based medicinals are available to control and actually cure many of the problems which the expensive, patented, brand-name, snake-oil currently being advertised on TV with tens of millions of dollars annually can't quite cure. Political shifts in the winds will eventually expose some of this. Don't wait, dump this industry as quick as you can in all forms. Sell all stock, refuse to buy the products, seek alternatives. And if you know a drug industry employee, tell them you are really disappointed in the whole lock stock and barrel of them.

Computers Unless You Are An Electronics Genius

Surely you are kidding. The Sun Apple PC Microsoft computer revolution has run its course. The stocks are still way overpriced. Neither value nor jobs will come back into this industry like before. Consolidation is the order of the day, and eventually it will come to pass that a whole new design solution will appear from radical upstarts, maybe from garages in places like Bharati (India), which will leave all the Apple/PC's of today sitting in Goodwill bins.

On the otherhand, roadside electronics/video/computer guru second-hand "keep it working" shops are likely to play an ever more important role during the coming years. Especially in rurals area.

Dot-Coms – Yuppie Style

Dot-Coms are a joke. Never invest. The problem is very simple. You cannot "own" the internet, you cannot get some sort of hegemony over it. All you can do is offer what you do. Offering to be a dot-com isn't it. To make money at it, you have to offer something unique, a true value-added magic or service which provides people with something they want or can enjoy.

Those who do unique things in the real world for real, can also make money through the internet, perhaps far more money than they could on their own without the internet. But it doesn't work the other way around because the typical dot-com doesn't offer anything which is unique, it doesn't offer enough value-added which will create a sale.

Amazon.com didn't become successful because it decided to sell books on the internet. It became successful because it offered a compelling service for bookworms which was not available anywhere else. It provided a look up service for any title ever produced. You could determine if the book was still available, what it's price was, how to find used book dealers, and information about the title and the author, all things you couldn't acquire conveniently. As a convenience, you could even order the book from Amazon.com That was the formula.

The internet is best conceived as a cooperative of producers who agree to share information in a way which makes the world "a transparent village" with the means for people to interact to learn things, find things, and complete transactions. There is very little role or room for the typical retail mentality, for the "middleman" trader, which is what most of the dot-coms tried to be. Nobody in the whole world needs another roadside squatter who thinks his " commercial real estate" is somehow special. Sorry, there is nothing special about any of them.

Energy

It is very difficult to generalize on this industry. In general, look for new energy tech which can be competitive at today's prices with oil and gas. There appear to a variety of new technology products streams which could heavily impact automobiles and trucking, even possibly farms, small business operations, and rural homesteads. Look for companies which are actually selling the new energy products or new ways to use energy more efficiently. Don't get involved at any other level as this is an arena which has been heavily played by scamsters throughout the past 50 years. Promises and possibilities don't work, only saleable products.

Financial Industry

Brokerages, Banks, and Insurance Companies will fall upon truly hard times. Many may go in bankruptcy. Paper equities, even insured equity in insured institutions, are not necessarily all that safe and secure. Equities, accounts, annuities, and other paper in this industry is going to be worth substantially less in 2008 than it was worth in 2003.

If annuities, life insurance, pension fund plans, special brokerage equity accounts, and other forms of paper involvments with the large financial services companies are important to your sense of well-being, you need to do a complete inventory of your life and begin to ask many serious what-if questions about whether and/or how those accounts and companies might fail during a general melt-down. You may need to do some serious hedging outside of financial instruments.

Gold & Silver & Non-food Commodities

Whatever you do don't get suckered by gold shills. At about $450/ounce, the supply can be expanded to accommodate ANY increase in demand. Same logic as with silver, but at much lower price. Silver simply is too easy to mine in vast quantities. As the Hunt brothers discovered a few decades ago, it is impossible to horde silver. The same can be said probably for all metals unless you have the inside tips on rare metals or rare earths which may be in particular demand by military or space systems.

If the base price in dollars for metals begins to noticeably appreciate , it is probably because the price of a barrel of oil is going up as well as the euro. If so, it is a total waste of effort to buy metals. Just buy an oil well. Or buy euros.

Health, Healing, Medicine, & Longevity

This is another area which is very difficult to generalize about. In general, the old new wave of huge bureaucratically managed conglomerates is OUT. There are many opportunities in many forms in this arena. Go with well tested items or with items which are easily proven directly to you. Refuse all mystification with "cure" products in which the ingredients are a secret. No matter how alluring such products may sound, all such products are based on a scam.

Mass Broadcast Media - American Style

The Broadcast Media has degenerated into the lowest common denominator, which is not a business you want to be in. A great portion of what is on the TV is neurotic, mentally diseased crap for brain-dead idiots. Advertising dollars will shrink radically, and numbers of channels may shrink. After the collapse, mindsets in most cultures will change dramatically, opening up new opportunities for new faces, new messages, new theatre, new relevancy.

Print Media

Advertising dollars will shrink radically, content will shrink, and jobs will shrink. Hard times for the newspapers but they will enjoy somewhat of a renaissance. More people will have more time to read and they will actually be desirous of reading to figure out how to avoid extreme personal stress and find new personal solutions.

Real Estate & Construction

In general, much real estate will be worth half as much in 2008 as it is worth in 2006 in many of the large urban areas. Some areas of the United States will continue to boom with new construction for second and third homes, such as along the Mogollon Rim in the Four Corners region, stretching from Cottonwood, Sedona and Flagstaff through the White Mountains into New Mexico all the way to Santa Fe. Other areas will stagnate and decline. Valuation deflation will hit all markets and all tiers and turn-over will drop to 10 percent of current levels. The incredibly shrinking Middle Class will shrink even more.

Retail Chains - Big Mass Retailers

The amount of Retail Space in many urban areas is absurd. The repetition of the same chain stores every few miles is crazy. Most values in this industry will shrink substantially this next few years and some portions of retail real estate will stand empty waiting to be adapted to new uses.

Software Unless You Are A Software Engineer

There are always new applications and new technology and new opportunities which can be integrated into new systems solutions. But this is mainly an insider's game. Don't get connected unless you want the software yourself and can see that third parties have very good things to say about it. Never listen to brokers about this industry. Microsoft is way past prime and its core software business has nowhere to go but down, which will nosedive severely for a few years. The company has so much money its ability to survive is probably not in doubt, but just how is an open question.

Travel & Tourism

Hard times for some regions, continued good times for others. Conditions are probably unpredictable in general and probably pessimistic for many venues. Terrorism has had tremendous negative impact on this industry. SARS was a devastating follow-on in certain areas. Business travel is not going to recover much, which will leave the service industry depressed. Tourism will provide some hope for some areas. Hawaii probably never will recover in this age of humanity. Yet notable scenic wonders, especially areas like the Grand Canyon and the Four Corners area continue to do well. Europeans have a tremendous fascination with this part of North America and the rising value of the euro will help them travel to the U.S. Make your picks very carefully with detailed insider knowledge.

Pick The Winners

Your criterions should focus on smaller entrepreneurial companies and on mutuals which specialize in smaller entrepreneurial companies. Stick to basic producers, small companies with lots of growth potential. The best companies will have a few years behind them and are now showing a profit but are priced at less than 15:1 price/earnings. If not, just wait. Sometimes start-ups are well worthwhile IF they have products which can be sold immediately.

Automotive – New Tech

Diesel hybrid, hydrogen, and electric propulsion systems are likely to go big time as part of a massive national industrial development program to get America's fat asses out of lawyer's offices or unemployment application benches. It is just a question of timing. The best you can do is position yourself. Not much this year, there will be more next year, and so on, and by the last two years of this first decade of Century 21, this area may begin to BOOM as politicians shovel truckloads of cash into the industry to create a lot of jobs quick. This could happen even earlier, perhaps as early as 2005 or 2006.

Aviation Civil Eurasian

Eurasian aviation will do a bigger intercontinental business in direct proportion to the falling value of the dollar. More and more tourists from abroad will be lured by cheaper and cheaper U.S. prices to come to visit. Could be a lot of growth in this area to absorb some of those thousands of acres of parked U.S. jets

Aviation Military

Since military aviation is the home page for a great portion of America's black arts programs, which involves money laundering for tens of billions of dollars of "secret money" for "black programs which do not exist", Boeing and related or like military suppliers will continue overall to do well despite the virtual collapse of civil aviation.

Cottage Manufactures - Consumer

It's a big world with a lot of needs. There is always a way for new innovation or just plain old standbys in the basic products needed by consumers. For the next few years, the best opportunities may be with the most basic items for which people will trade regardless of the larger economy.

Dot Coms – Country Style

Think of the most remarkable specialty stores in out-of-the-way places you have ever been in. They are endlessly fascinating because you can typically learn a lot about a specific topic. Nerds in Silicon Valley can't do this stuff because they just don't think along these lines. Their primary focus is in computers and software and mass markets for simple minded widgets and ideas. So "specialty" competition is limited. That's how to do the internet.

Electronics

There remain a stunning number of potential product concepts in the electronics field. Expand your vision beyond computers and digital. Many "unique" tech products can be made on a cottage industry basis.

Energy

See the Loser's List – there may be very good choices in this field but be careful, it is a highly competitive industry in some respects, highly rigged in others. Eventually, money will be shoveled into this industry to buy new production technology which creates energy far closer to its end use.

Eurasia

Eurasia is a big place and has many huge markets. Smaller companies which are doing well in Eurasia with an expanding market are excellent prospects. What could be made and exported to Eurasia if the value of the dollar falls another 10%? 20%? 40%? Can it remain in demand even if European economies enter into a serious depression?

Euros

The euro will continually appreciate vis a vis the Pound and the Dollar. Why not save your money in a euro account, even possibly in a European bank?

Food – Agriculture

This is an area which is ripe for a major paradigm shift. More intensive production, in smaller units, using much more robotics, will eventually come into play. This would be a very fruitful area for a National Industrial Development Program.

Health Field

This area will experience a profound paradigm shift. (See the Loser's List). The proponents of "healing' modalities are now sophisticated and popular enough to overcome some of the high cost invasive medical practices. As a part of this paradigm shift, provable protocols for relatively inexpensive cures for cancer, arthritis, diabetes, and Alzheimer's diseases are likely to emerge during the next several years. Removing the burdens of these widespread degenerations from the health care field will enable a universal health care system to be implemented at far less cost than currently supposed. Re-engineering the entire field should be one of the primary areas of a national industrial development program.

Materials – Advanced

Where's "starlight", a new polymer invented in Britain many years ago which is reputed to block electromagnetic energy, heat, and radioactive particles? It disappeared down into a huge gaping maw of the black arts empire of the military-industrial complex. There are many potential new industries in this area, but military competition has hidden it from us. Perhaps the policy makers of a national industrial development plan can bring this and many others to the fore.

New Media

New cable and satellite channels will appear along with new internet media operations. Aim for planetary perspectives. Everybody everywhere is weary and wary of the old nationalism.

Psychic Phenomenon

Intriguingly, in September 1939, Cayce predicted that spiritual phenomenon (astral/etheric plane consciousness) COULD be turned into as reliable a science and as practical a technology as any other, but he made that conditional on the application of as much resource to develop it as had been given the other sciences of the industrial age.

Purification

Human beings are poisoning themselves at many levels. Product streams which purify food, water, air, fuel, buildings, even media, are badly needed. New technology development is perpetually redesigning the art of the possible and generally the commercial applications lag far beyond what is possible. Renewed diligence and investment in purity will be a part of the new National Economic Industrial Policy.

Real Estate & Construction

See the Loser's list. And consider. Do we really want to live and work in most of the current inventory of buildings in the world? There are many reasons related to energy and health for supposing that this industry could be a boom industry after the recovery.

Remote Sensing

Environmental monitoring, transmission, reception, and display can accomplish far more tasks than they currently are. The focus on security has generated a lot of technology and services but the applications are still primitive in most regards. Think way out of the box. Combine this with

Robotics, below, for one of the primary zones of the new National Economic Industrial Policy.

Roadside

Don't laugh. It works. A lot of people during the Great Depression got into selling door to door, even apples.

Robotics

This is probably one of the primary major growth areas for the new National Economic Industrial Policy. The PC industry for "information" is saturated. Cyber-mechanics is next. Robots "in the walls" for cooking, health monitoring, auto maintenance, specialty crop management, retail distribution, there are enormous potentials in this field.

Rural Small Towns

The huge urban megaplexes are just too uncomfortable and unhealthy to live in unless you can spend a small fortune isolating yourself in special "elite" environments. Think seriously about the hinterland and small towns. There is a lot of hinterland, small towns are everywhere, costs are generally lower in them, and people with families always prefer the easier pace of life and the higher quality environment. Generally competition is not as great and there are always lags in the hinterland in the supply of goods and services, thus there are many more opportunities than might be supposed. Be sure to pick a growing region, not a dying one.

Security

The brawn and technolgy of this industry is over-extended. Think about brains for this industry.

Sunspot Cycle Charts

Note: color versions of these and other sunspot graphs and charts are available on the internet: michaelmandeville.com/earthmonitor/cosmos/sun/sunspots.htm

Chart 101: Sunspot Cycles & Human History

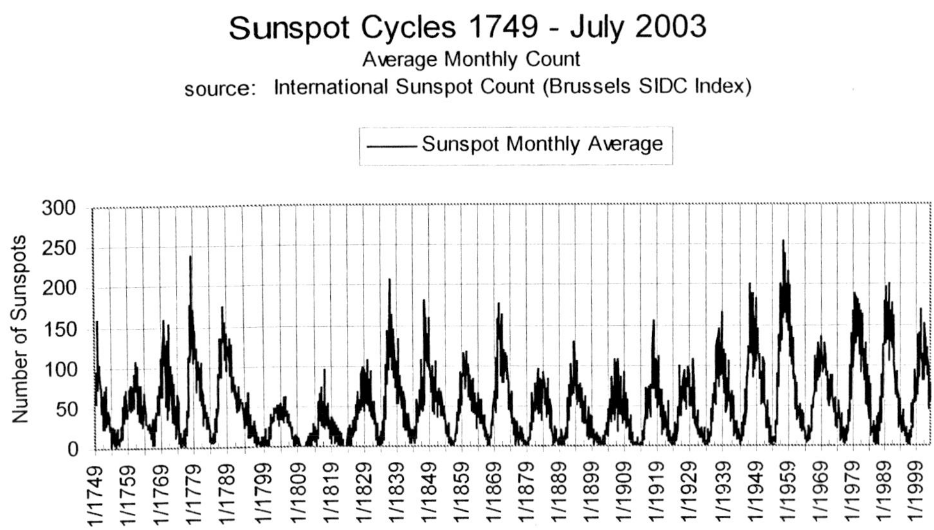

Compiled from the International Sunspot Count by MWM, 2002. Most major human conflicts are easily correlated to the peaks in the count of sunspots and so are major weather disturbances, including severe droughts and floods. MichaelMandeville.com

What Are Sun Spots:

Sunspots are seen as "small" dark spots on the surface of the sun. They are easy to observe and count if the sunlight is strongly filtered. They were first noticed (in Western record) in the year 325 BC by Theophrastus, an Hellenic scientist, and they have been counted on a regular basis since the middle of the 17th century. They come and go in cycles which average about 11 years, as shown in Chart 101 above.

These "small" dark spots are conceived by modern astrophysicists to be intense "bubbles" of magnetic energy which somehow cool down the hot gasses within so that they appear dark compared with the surrounding solar atmosphere. These "cool" bubbles are not really very small, they are quite often the size of the Earth and many times giant spots many times the Earth can be seen. Many more physical facts about them can be can be found on the NASA and NOAA websites, such as at http://www.sec.noaa.gov/primer/primer.html

A. L. Tchijevsky, a Russian professor of Astronomy and Biological Physics, noticed during World War I that particularly severe battles followed solar flares. Since the sunspots were in a peak period during 1916-17, no doubt the war and its various battles were heavily stimulated by the energies which are boiling off the Sun. Intrigued by the connection of human behavior to solar physics, Tchijevsky constructed an "Index of Mass Human Excitability". He compiled the histories of 72 countries from 500 BC to 1922 AD to provide a strong database to articulate his correlations. After rating the most significant events, Tchijevsky found that fully 80% of the most significant human events, mostly related to war and violence, occurred during the 5 years or so of maximum sunspot activity.

Tchijevsky went on to observe that the 1917 Russian Revolution occurred during the height of Sunspot Cycle. Unfortunately, this was one of science's most costly observations, it earned Tchijevsky almost 30 years in Soviet prisons because his theory challenged "Marxist dialectics".

The "solar" connection to terrestrial events has been studied ever since then, but most of the focus has been on the sun itself or on the impact of the cycle on the climate, weather, agriculture, commodity markets, and other non-human phenomenon. Awareness of the human impact, which is far more significant than the well known impact of the Full Moon, has remained highly retarded. Modern humans, unlike the ancient cultures of Egypt, Sumer, Bhararti, Maya, and China, are highly reluctant to admit that their collective behavior is influenced strongly by the Sun. They prefer to *believe* that *reason* rules their societies.

Chart 101, above, was compiled by compressing all of the average monthly sunspot counts for the past 254 years into this simple graph to show the full range of variation in the average monthly number of sunspots. 23 distinct cycles are shown here, beginning with a peak year in 1749. This chart, and all the others used in this Section, are based on what is called the "ISSN", which is an acronym for the "International Sun Spot Number" which is the consensus count made by observatories every day.

As can be seen in the chart, there is a great deal of variation in the average monthly counts and these in turn make quite a variation in the size and width of the 23 sunspot cycles. Note that there are three sunspot cycle peaks which did not have monthly peaks in excess of 100 and there were at least five with monthly peaks which reached 250 or more. That is quite a range for a dynamic cycle and we should expect that the effects in the solar system and in the Earth will show a similar variation. Most likely "the shadow" of the solar cycles can be readily seen in thousands of chemical, mineral, biological, and economic data series which scientists make by studying plants, minerals, and human history. And most likely "the shadow" varies considerably.

Data Sources:

The official International Sunspot Number, which is also known by NOAA as "RI", is issued by the Sunspot Index Data Center (SIDC) in Brussels. The ISSN comes in three flavors, a daily count, a monthly average, and a yearly average. You can also use "smoothed" numbers, which round off the numbers. Astrophysicists may have a use for smoothed numbers, but for connecting the Sun with weather and human events, the simple counts and averages are generally far more appropriate. Data and plots are available from the SIDC web site at http://sidc.oma.be or at the NOAA website http://www.ngdc.noaa.gov/stp/SOLAR/getdata.html

Detailed graphs of each sunspot cycle can be found at John Alvestad's website at http://www.dxlc.com/solar/ These can be used for paralleling events with sunspot

peaks. A great quantity of explanatory material and various sunspot numbers are provided by NASA and this webpage is useful for explaining sunspot numbers: http://science.msfc.nasa.gov/ssl/pad/solar/greenwch/spot_num.txt
A huge, ultra-wide sunspot chart which shows each year since 1749 in complete clarity can be found at
http://www.michaelmandeville.com/earthmonitor/cosmos/solarwind/sunspots.htm

History Cycle Table

The following History Cycle Table is based on the average annual ISSN sunspot number. For defining "peak periods", one has to set a "bar" for what constitutes a "peak". Is it 100? if so, we lose several cycles. If it is 50, we gain them all, but the periods are "fat" and include a lot of years in between years which may not be very significant. Most likely, to see the validity of the connection of sunspot peaks with human violence, it is best to set the bar "high" to narrow the number of years. If the major wars all fall within these limited number of narrow bands, it is clear that the connection is very real.

To make sure we include all sunspot cycles, the year of the maximum average sunspot count in every cycle was used to define the high point of the solar cycle. The year before and the year after it are added to define the "nominal' sunspot peak years. For the sunspot cycles with high counts over a longer period of time, a "bar" was set at 100 and all years which were above 100 were included in the "peak" for that period.

This is an arbitrary method because the sunspots vary considerably in peak size and also because each of them has a somewhat different peak width. In truth, it is hard to generalize specifically about a sunspot peak because of the high degree of individuality and variability which they show.

This method was adopted mainly because of the thesis that human reactions are not caused so much by the absolute numbers of sunspots, but by substantial "changes" in the numbers which drive "shifts" in human mental and emotional processes By using this "dual" method for defining a peak , we have a simple, convenient way to define the periods of maximum change regardless of the numbers.

The information in the following table is of course merely illustrative. The number of wars and major economic events which "connect" with the sunspot cycle peaks are much larger than the small number of "major events" which are included here. I have omitted data for the first seven earliest cycles. Prior to the first named solar cycle, which is called Solar Cycle 1, there were at least five cycles for which there is a good consistent profile of daily counts. For students: filling these in with your own school reading and learning assignments would make a first class term paper.

Solar Cycle 1704-1706
Solar Cycle 1716-1718
Solar Cycle 1726-1728
Solar Cycle 1737-1739
Solar Cycle 1749-1751
Solar Cycle 1: 1760-1762
Solar Cycle 2: 1768-1770

Solar Cycle 3: 1777-1779

1776-1783	American Revolution

Solar Cycle 4: 1786-1788

1788-1791	French Revolution
1789	US Constitution adopted

Solar Cycle 5: 1803-1805 under 100 wide (1802-1806)

1803-1806	Napoleon conquers Europe

Solar Cycle 6: 1815-1817 under 100

1815-1817 Two wars to defeat Napoleon; German, English and Serbian riots; Brazil, Chile and Argentina declare independence.

Solar Cycle 7: 1829-1831 under 100

1828-1832 Revolts in Turkey, Mexico, Belgium, Poland, France, Britain; Virginia slave revolt; the Black underground railroad begins

Solar Cycle 8: 1836-1838

1837-1840 Constitutional revolts in Canada, slavery debate outlawed in US, Texas Independence, Boer separatists occupy African lands, British-Afghan war; Opium War

1937 Major Banking Crisis in the U.S.

Solar Cycle 9: 1847-1849

1846-1848 Mexican War

1848-1851 Revolts and revolutions in Poland, Switzerland, Paris, Vienna, Berlin, Milan, Venice, Naples, Prague, Budapest, Warsaw; US Mexican War starts; Taiping Rebellion starts

Solar Cycle 10: 1859-1861 under 100

1858 Bottom year of a depression in the U.S.

1861 American Civil War begins

1861-1865 Civil War in America, revolts in India, Italy, China

Solar Cycle 11: 1869-1871 wide 1869-1872

1869-1870 Franco/Prussian War

1869-1872 Paris Revolutionary Commune

Solar Cycle 12: 1882-1884 under 100

1883-1886 Big US labor strikes, revolt in Sudan, First Indian Congress meets

1883 Bottom year of a major depression in the U.S.

Solar Cycle 13: 1892-1894 under 100

1893-1895 Zulu revolt, Cuban revolution

Solar Cycle 14: 1905-1907 under 100

1904-1905 Russo-Japanese War

1905-1908 first revolts begin in Russia

1905-1908 Widespread strikes, revolts among German miners, Hottentots, Turks, Indians, Honduras

1908 Bottom year of a short depression

Solar Cycle 15: 1916-1918 just barely 100

1914-1918 First World War

1916-18 Irish and Indian revolts

1917 Russian Revolution

1919 The Atom is Split

Solar Cycle 16: 1927-1929 under 100 wide 1926-1929

1927-1929 Fabled American Bull Run ends in crash of the stock market in long slow slide which bottoms in 1933

1926 Hitler in jail for NAZI's attempted Munich Putsch, begins writing Mein Kampf which outlines how he will lead Germany to make the world's greatest power.

1927-1931 Mussolini and Hitler build power on economic unrest; revolt in Vienna and China; formation of Red Army; Spanish Republic formed; mass civil disobedience in India launches Ghandi's campaign to free India

Solar Cycle 17: 1936-1938 wide 1936-1939

1936-1939 Spanish Civil War, Germany and Japan start World War II

1937-1940 US steel strike

Solar Cycle 18: 1947-1949 wide 1947-1950

1946-1949 Greek Civil War, India-Pakistan riots, Red Army wins China, Vietnam revolts

1947 - 1948 Flying saucer sightings begin, saucer crashes in Roswell, NM, "shadow" government is set up inside the military industrial complex with the CIA to

	fight communism and hide the remains of ET
1948	Ghandi assassinated, Israel's War for Independence
1950-1953	Korean War

Solar Cycle 19: 1956-1958 huge peak wide 1956-60

1957 - 1960	Israel invades Sinai, Hungarian uprising, Cuban revolution, civil rights movement begins in US, French-Algerian war, MauMau revolt, Iraq revolt,
1957	Vietnam War begins
1958	Eisenhower recession
1960-1961	Eisenhower warns of the danger of "shadows" in the unfettered military industrial complex, Kennedy "race to the moon" begins

Solar Cycle 20: 1967-1969 stumpy wide 1967-1970

1965- 1967	Haight-Ashbury Flower Children launch the hippie movement
1967-69	Height of Vietnam War, peace demonstrations, worldwide student uprisings, Czechoslovakian uprising/USSR invasion, US inner city riots, Israeli Arab war, Woodstock and height of hippy movement,
1968	Martin Luther King and Bobby Kennedy assassinated , first big anti-war marches in US, first US inner city riots,
1969	First public men on Moon

Solar Cycle 21: 1978-1980 wide 1978-1982

1978	World's First Test Tube Baby Born, Carter Camp David Accords between Israel and Egypt
1979	Three Mile Island Nuclear Plant leaks radiation
1979-82	Polish Solidarity begins, US anti-tax movement reaches heights and elects Reagan, Shah of Iran overthrown, Iraq-Iran war begins, USSR invades Afghanistan, Falklands War, Sandinistas oust Somas, Zimbabwe gains independence, anti-nuclear and peace demonstrations increase worldwide, US aid to "contras" in Nicaragua, US invades Grenada, Tamils rebel in Sri Lanka
1979-1980	US Bid to Rescue Hostages Fails
1980	Iran-Iraq War begins (lasts until next peak in 1988)
1981	President. Sadat of Egypt assassinated
1982:	Israel Invades S Lebanon, Falklands War
1982	Reagan recession

Solar Cycle 22: 1988-1990 wide 1988-1992

1987-88	Palestinian Infiltada begins, Eastern European dissidents organize, USSR & US sign missile treaty, USSR begins pullout from Afghanistan
1987	Crash in World Stock Markets
1989	Tianamen Square Chinese student democracy movement crushed
1989	Protest and peaceful revolution in Eastern Bloc, dismantling of Berlin Wall,
1989-1992	Glasnost process begins dissolution of Soviet Union, end of Communist Party domination; Communist Party coup in Russia fails
1989-91	End of apartheid in South Africa, beginnings of patriot and militia movements in US, Somalia civil war, Yugoslavia begins slaughter in Bosnia, Sandinistas lose Nicaraguan elections
1990	Mandela Released, East and West Germany Re-Unite
1990	Iraqi Troops Invade Kuwait
1991	The Gulf War; multi-national forces liberate Kuwait from Iraq, Balkan Civil War begins as communist Yugoslavia collapses.

Solar Cycle 23: 1999-2001 wide 1999-2002

1998-2000	Peace treaty in Northern Ireland, overthrow of Indonesia's Suharto, Serbian-KLA conflicts increase and US/ NATO decides to "resolve" conflict through massive bombing of the whole nation, India-Pakistan skirmishes over Kashmir increase; militias burn East Timor, drive people into camps in West Timor; big demonstrations at WTO meeting in Seattle; rising religious strife in India and Indonesia, increasing civil war in Sierra Leone and Sri Lanka; overthrow of Milosevic in Serbia;
1999	Worldwide Y2K Scare
1999	Palestinian Infiltada re-commences after "virtual agreement" with Israel

2000	Dot-com bubble breaks; Supreme Court intrudes in the U.S. elections, throws results of Florida vote to make the loser of the popular vote into a winner of the electoral vote to become the U.S. President made by judicial interference.
2001	Attack on World Trade Center and Pentagon; War on Terrorism begins
2000-02	Peace treaty disrupted in Northern Ireland; land confiscations in Zimbabwe; worldwide protests against WTO/IMF/World Bank in Seattle, Washington DC, Prague, Goteborg, Quebec City, Genoa and other cities
2002-2003	Bush diverts War on Terrorism into personal vendetta against Saddam Hussein, massive protests demonstrate against Iraq war

Chart 102: Sunspots Cycles & Last Two Recessions

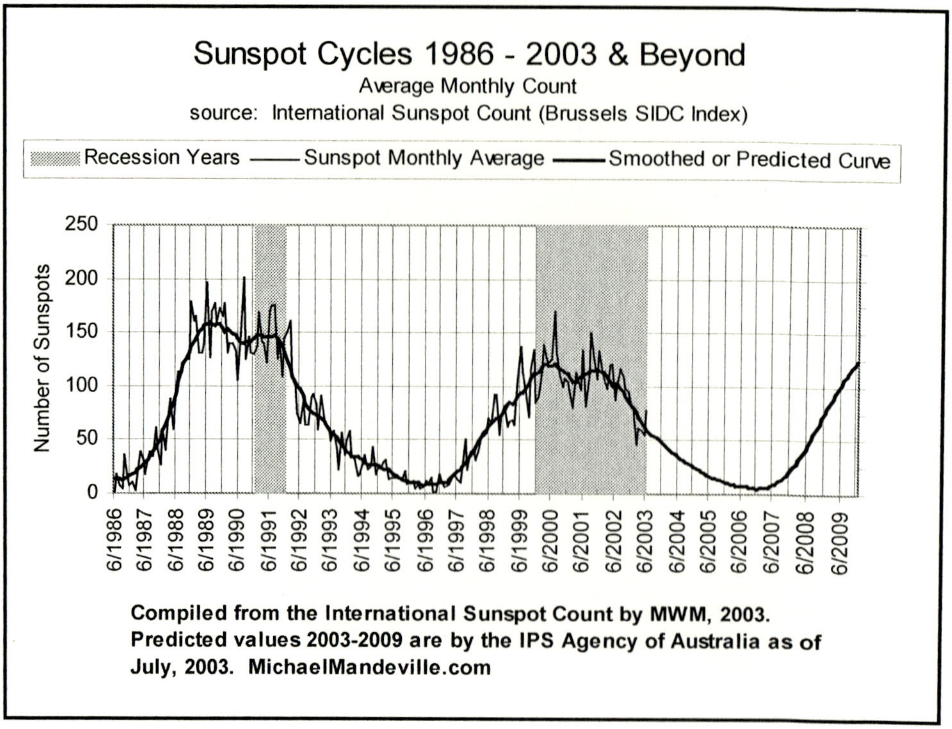

In the History Cycle Table, it is easy to see that both political and economic affairs are profoundly caught up and influenced by the "waves" of sunspot energy. The connections are even easier to spot if we see the sunspot cycle in greater detail. In Chart 102 above, just two cycles are displayed, the current one out of which we are gradually emerging and the last one which took the world on a highly transformative ride from 1988 to 1992.

The monthly average counts are shown in the jiggly line and the "smoothed" average annual curve is shown to better define the overall cycle. As we can see from this chart, the first "Bush" recession in 1991 came during the last part of Cycle 22. It was preceded, of course, by the 1987 crash in world stock markets which came just as Cycle 23 was on its way up.

The most recent "bubble" crash in 2000 initiated a recession which began almost immediately. Though Republican propagandists are now trying to convince people that the recession ended in 2001, massive job losses through to 2003 suggest that the recession did not end until, maybe, sometime during the early part of 2003 year. It remains arguable in July 2003 that the recession is in fact not yet over.

Chart 103: Sunspot Cycles & Major Economic Contractions 1926-2003

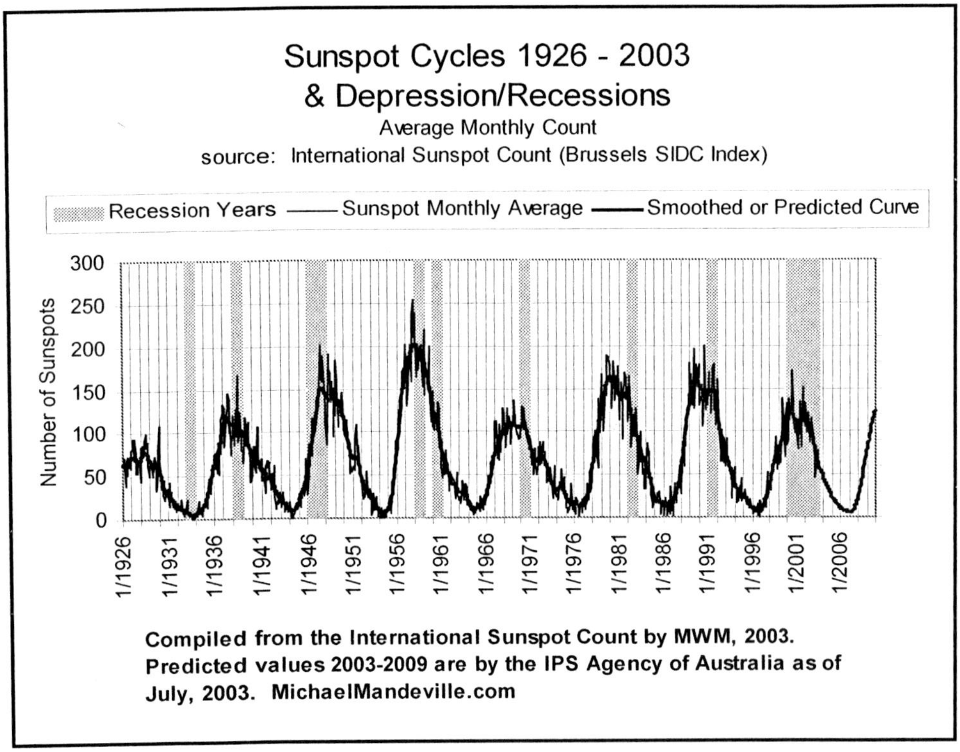

How strong is this historical connection between major economic downturns and the sunspot cycles? We can learn more about this connection of sunspots to economic downturns by directly graphing them together in Chart 103, above. Quite clearly, Chart 103 shows us that there is a rather strong connection between major recessions and the peaks of the sunspot cycles. There was one major exception, the last Great Depression, the bottom year of which (1933) can be seen in the trough between sunspot peaks. The next Great Depression may parallel this exception nearly to a T.

Can this historical connection be used to predict stock prices? Is there a correlation between sunspot peaks and the Dow Jones Industrials? (Dow Jones

Industrials: these are select stock prices often referred to as the DJI) There is no usable connection except as a harbinger of a coming break. There is zero correlation between daily price movements and average daily sunspot numbers. Is there a connection between long term historical trends in the prices and average monthly or annual trends in the numbers of the sunspots? Not really, the only direct connection that appears is as a "breaking" signal. During a sunspot peak, the speculative Bull Run bubbles in stocks "break" and an economic recession begins fairly soon thereafter. This often leaves the stock prices headed down even while sunspots are still rising. This destroys any statistical averages which can be used for prediction.

As can be seen in the graph, there is a decidedly strong parallel between recessions and the peaks. It has been consistent throughout most of the century with one notable exception. The bottom year of the Great Depression in 1932/33 was at the bottom of the sunspot cycle. The collapse of the stock market, however, paralleled right on the peak of the solar cycle in late 1929. Stock prices slid as sunspot numbers slid, and the economy wallowed as sunspot counts reached 0.

We may be paralleling the 1929 to 1933 era. There is probably a strong tendency in this era to continue to slide after the bubble break in 2000 for a few years until all of the speculation has been squeezed out of stock prices. From the Bears, we already know that this means stock prices generally must fall yet another 35% to 75% from their levels in June 2003, depending upon the industry and the company.

This will eventually probably be the outcome of the current 25 year long depression cycle and we are likely to catch up with this inevitability in 2006 and 2007. In the meantime, most likely we are currently still buoyed up by massive subsidy stimulation, 70 years of institutional barriers, and various social security buffers. This may be enough to create a very modest "faux" bubble amidst a "jobless" economic recovery.

Bibliography & Notes

Section One – What This Report Is About

"*25 Year Economic Depression Cycle*" by Edgar Cayce, see "*Return of the Phoenix Book Two: The Great Breakup*" by MW Mandeville; MetaSyn Media 1999; p. 181

Mandeville, Michael Wells: "*Return of the Phoenix: A Trilogy In Three Volumes*"; Black Canyon Arizona: MetaSyn Media, 1999.

Mandeville, Michael Wells: "*Return of the Phoenix: Book One – The Veil*"; Black Canyon Arizona: MetaSyn Media, 1999.

Mandeville, Michael Wells: "*Return of the Phoenix: Book Two – The Great Break-Up*"; Black Canyon Arizona: MetaSyn Media, 1999

Mandeville, Michael Wells: "*Return of the Phoenix: Book Three – The Prophecies*"; Black Canyon Arizona: MetaSyn Media, 1999.

Mandeville, Michael Wells: "*Return of the Phoenix CD-ROM*"; All three volumes of the Trilogy are included in four versions: (1) elaborately illustrated in html form for reading with any web browser, (2) in Acrobat PDF documents, (3) in MS Word 7.0 (PC) documents, and (4) as MS Word For Macintosh 5.1 documents; Black Canyon Arizona: MetaSyn Media, 1999. Also available on the Iway at MichaelMandeville.com/

Sunspot Cycles, see
http://www.rexresearch.com/prophist/phf6cy~1.htm
http://www.borderlands.com/sun/sunspots.htm
http://www.tidesofchange.org/geo.sun.htm

Section Two – The Coming Economic Collapse of 2006

Accuracy of Cayce's objective historical and science predictions in the "*Return of the Phoenix: Book Two – The Great Break-Up*" by Michael Wells Mandeville; Black Canyon Arizona: MetaSyn Media, 1999; see the first four chapters.

"*25 Year Economic Depression Cycle*" by Edgar Cayce, 09/08/31, Reading No. 282-003; see "*Return of the Phoenix Book Two: The Great Breakup*" by MW Mandeville; MetaSyn Media 1999; p. 181

Cayce/Davis Collection on CDROM; A.R.E. Press, 1992; sold under the title of "The Edgar Cayce Readings"

Earth Changes Almanac & Calendar, 2003+, Black Canyon Arizona: MetaSyn Media, 2003; online html documents at http://www.michaelmandeville.com/calendar/

Mandeville, Michael Wells: "*The Coming Economic Collapse of 2006*"; Black Canyon Arizona: MetaSyn Media, 2003; PDF ebook
http://www.michaelmandeville.com/calendar/

Mandeville, Michael Wells: "*Transmutations*"; Black Canyon Arizona: MetaSyn Media, 2003; PDF ebook in progress due late 2003.
http://www.michaelmandeville.com/calendar/

385 prediction by Edgar Cayce on 10/19/28: No. 900-412 in the "*Cayce/Davis Collection on CDROM*"; see Mandeville, Michael Wells: "*Return of the Phoenix: Book Two – The Great Break-Up*"; Black Canyon Arizona: MetaSyn Media, 1999, see 900-412 on 10/19/28

Vortex Tectonics, on CDROM, by Michael Mandeville, Black Canyon Arizona: MetaSyn Media, 2001

"*World Business Cycles*"; London, Economist Newspaper Ltd.; Detroit: Gale Research, 1982.

Section Three – The Global Political Shift

BBC News – The population of Palestine is not known from an census. BBC News estimates it to be 2.8 million. See:
http://news.bbc.co.uk/1/low/world/middle_east/country_profiles/803257.stm

Imperial Faction - some links:
http://www.encounterbooks.com/main_web/resources/resources.html

Internet News Media: http://www.globalpolicy.org/

PROJECT FOR THE NEW AMERICAN CENTURY
http://www.newamericancentury.org/
1150 Seventeenth Street, N.W., Suite 510, Washington, D.C. 20036
William Kristol, Executive Director

REBUILDING AMERICA'S DEFENSES: Strategy, Forces and Resources For a New Century; A Report of The Project for the New American Century, September 2000; Thomas Donnelly Principal Author, Donald Kagan & Gary Schmitt, Project Co-Chairmen; available in PDF format; see especially Page 9.
http://www.newamericancentury.org/

Regular, Arnon: "*Road map is a life saver for us, PM Abbas tells Hamas*"; Harratz, Wednesday, July 02, 2003 Tamuz 2, 5763 (see Harratz.com (an Israeli newspaper)

Rense, Jeff: national radio talk show host; Rense.com – extensive internet archive of professional critical news and magazine articles from press in many countries, plus free lance reportage from around the world on a wide variety of issues and cultural interests.

The Old Hopi, 2003
as quoted in "*Transmutations*", by MW Mandeville, 2003

Thomas, William: "*All Fall Down: The Politics of Terror and Mass Persuasion and Bringing The War Home*", 2003, http://www.willthomas.net/ or willthomas@telus.net

Thomas, William: "*The Prozac War*", March 6, 2003
http://groups.yahoo.com/group/earthchange-bulletins/message/1113

Thomas, William:, "*Is Bush Nuts*", 2003
http://www.willthomas.net/isbushnuts.htm

Weekly Standard, The; For a review of the nuttiness in the very center of the wacked out wight wing, see http://www.weeklystandard.com/

Zionism, Jews, & the PNAC: "*Whose War? A Neoconservative Clique Seeks To Ensnare Our Country In A Series Of Wars That Are Not In America's Interest*" by Patrick Buchanan; The American Conservative, March 24, 2003
http://groups.yahoo.com/group/earthchange-bulletins/message/1164

Section Four – The Global Media Shift

Gilbert, Gustave: "*Nuremberg Diary*"; Farrar, Straus and Company, 1947, pp. 278-279; "*Mass Fear & War – How to impel a country to war*"; interview of Hermann Goering

Goering, Hermann, interview by Gilbert, Gustave titled "*Mass Fear & War – How to impel a country to war*" in "*Nuremberg Diary*"; Farrar, Straus and Company, 1947, pp. 278-279

McLuhan, Marshall and Fiore, Quentin: "*The Medium is the Massage*"; co-ordinated by Jerome Agel; Bantam Books / Random House, 1967; republished by Gingo, 2000
http://www.marshallmcluhan.com/pub.html
http://www.gingkopress.com/

McLuhan, Marshall and Powers, Bruce: "*The Global Village: Transformations in World Life and Media in the 21st Century*"; Oxford University Press, 1988.

Shirer, William Lawrence: "*The Rise And Fall Of The Third Reich; A History Of Nazi Germany*"; Simon and Schuster, 1960.

Shirer, William Lawrence: "*The Rise And Fall Of The Third Reich; Part 2, Nazi Germany, Years Of Triumph*" - a videorecording; MGM presentation produced by Wolper Productions. This is a Videocassette release of a 1966 motion picture which was made into a series of three films versioned by Films Inc., 1971; traces the expansion of Nazi Germany from Hitler's accession as Chancellor to the fall of France, shows how terror and propaganda can transform a democratic government into an absolute dictatorship, examines the goals and ideals of the Nazi state.

Section Five – The Global Economic Shift

For daily discussions of economic variables, numbers, and trends, see

"*John Mauldin Investors Insight*" email newsletter by John Mauldin; inquire <johnmauldin@investorsinsight.com>; offers high quality statistical summaries from a variety of sources and up-to-date arguments on the trends.

"*The Daily Reckoning*" email newsletter by Agora Financial Publishing; dailyreckoning@agora-inc.com; see http://www.agora-inc.com/reports/

For discussion of a key to a Century 21 National Economic Industrial Policy, see

Birch, David: *"Job Creation in America: How Our Smallest Companies Put the Most People to Work";* Free Press, October 1987. Birch has studied entrepreneurial activity since the mid-1970s at MIT; check out the following websites for additional information:
http://www.hi.is/~joner/eaps/cq_gazel.htm
http://www.cogonline.com/

Section Six – The Coming Dollar Panics

Dollar/Euro Trends: see

http://www.fxstreet.com/nou/noticies/senseframesnoticiaseurogran.asp?pv_noticia=/nou/euro/senseframeseuro2003526814.asp

Platt's Energy Advisory: http://www.platts.com/risk_management/futures/index.shtml

Halifax Herald Limited, The: Tuesday, May 20, 2003
http://www.herald.ns.ca/stories/2003/05/20/pBusiness157.raw.html

Economic Data about job, capital, and production transfers to Asian countries, see

McCarthy, John C.: "*3.3 Million US Services Jobs To Go Offshore*", Forrester Research, Brief of NOVEMBER 11, 2002; inquire press@forrester.com
http://www.forrester.com/

Petrodollar essays, see

Putland, Dr Gavin R.:, March 2003; Queensland University of Technology, Australia, see
http://www.fair.org/press-releases/venezuela-editorials.html
http://www.observer.co.uk/international/story/0,6903,688071,00.html

Heard, Geoffrey, March 2003, Melbourne, Australia, inquire gheard@surf.net.au.

New York Times, The: "*Dollar Could Careen Even Lower*", Thursday, May 15, 2003; http://www.nytimes.com

General Citations or References

Asahi Shimbun, The: "*Common Global Enemy - G-8 Needs To Coordinate A Fight Against Deflation*", an editorial, May 30(IHT/Asahi: May 31,2003)
http://www.asahi.com/english/op-ed/K2003053100189.html

Clark, W.: " *'The Real Reasons for the Upcoming War With Iraq: A Macroeconomic and Geostrategic Analysis of the Unspoken Truth*", January 2003 (revised 20 February),
http://www.ratical.org/ratville/CAH/RRiraqWar.html
or Independent Media Center, http://www.indymedia.org

FEASTA: The Foundation for the Economics of Sustainability: "*This War Is About More Than Oil. OIL DOLLARS!!!! DOLLARS, THE EURO AND WAR IN IRAQ.*" *February 10, 2003; see* http://www.indymedia.ie/cgi-bin/newswire.cgi?id=28334

Liu, Henry C K: "*Global Economy US Dollar Hegemony Has Got To Go*", Asia Times Online; http://www.atimes.com/global-econ/DD11Dj01.html

Sartre, "*The Petro-Dollar And The Euro Money Is The Root Of War*", Ether Zone, April 4, 2003; see http://www.etherzone.com/2003/kirk060302.shtml

VHeadline.com: "*USA Intelligence Agencies Revealed In Plot To Oust Venezuela's President*", *Thursday December 12, 2000*
see http://sf.indymedia.org/news/2002/12/1550023_comment.php#1551138

Vidal, Gore: "*The Enemy Within*", The Observer, London, Sunday 27 October 2002 http://www.ratical.org/ratville/CAH/EnemyWithin.html

Weisbrot, Mark: "*Split Screen In Strike-Torn Venezuela*", Washington Post, Sunday, January 12, 2003; Page B04
http://www.washingtonpost.com/ac2/wp-dyn?pagename=article&node=&contentI d=A41444-2003Jan11&notFound=true

Section Seven – Surviving The Collapse of 2006

For citations to Cayce's comments, *see "Return of the Phoenix Book Two: The Great Break-Up*". Look in the summary chapter Or do keyword phrase searches on the CDROM of the Cayce/Davis Collection.

Index

Printed in the United States
42801LVS00007B/25-30

9 780972 105194